EVERYDAY ENVIRONMENTALISM

EVERYDAY ENVIRONMENTALISM

Creating an Urban Political Ecology

Alex Loftus

 University of Minnesota Press
Minneapolis
London

An earlier version of chapter 4 originally appeared as Alex Loftus and Fiona Lumsden, "Reworking Hegemony in the Urban Waterscape," *Transactions of the Institute of British Geographers* 33, no. 1 (2008): 109–26.

Copyright 2012 by the Regents of the University of Minnesota

All rights reserved. No part of this publication may be reproduced, stored in a retrieval system, or transmitted, in any form or by any means, electronic, mechanical, photocopying, recording, or otherwise, without the prior written permission of the publisher.

Published by the University of Minnesota Press
111 Third Avenue South, Suite 290
Minneapolis, MN 55401-2520
http://www.upress.umn.edu

Library of Congress Cataloging-in-Publication Data

Loftus, Alex.
  Everyday environmentalism : creating an urban political ecology / Alex Loftus.
  Includes bibliographical references and index.
  ISBN 978-0-8166-6571-6 (hc : alk. paper)
  ISBN 978-0-8166-6572-3 (pb : alk. paper)
  1. Urban ecology (Sociology). 2. Political ecology. 3. Environmentalism. I. Title.
  HT241.L65 2012
  307.76—dc23
                                                2011047416

Printed in the United States of America on acid-free paper

The University of Minnesota is an equal-opportunity educator and employer.

Contents

ACKNOWLEDGMENTS   vii

INTRODUCTION
Emerging Moments in an Urban Political Ecology   ix

1. The Urbanization of Nature
   Neil Smith and Posthumanist Controversies   1

2. Sensuous Socio-Natures
   The Concept of Nature in Marx   21

3. Cyborg Consciousness
   Questioning the Dialectics of Nature in Lukács   45

4. When Theory Becomes a Material Force
   Gramsci's Conjunctural Natures   75

5. Cultural Praxis as the Production of Nature
   Lefebvrean Natures   109

CONCLUSION
The Nature of Everyday Life   131

NOTES   137
INDEX   161

Acknowledgments

Many of the ideas for this book began germinating at Queen's University in Canada among the revolutionary dreams of my friends Spencer Tracy, Tony Weis, and Matt Silburn. At around the same time, David McDonald inspired an interest in South Africa and water. Research in Buenos Aires was made possible and enlivened by Itziar Gómez Carrasco. Moving to Oxford, the frenetic pace of Erik Swyngedouw's wild and radical imagination was balanced by the steadying hand of Tony Lemon. To Erik, in particular, I am indebted for a large number of the ideas here—and also for opening up exchanges or friendships with so many others, too numerous to mention. In Oxford I met Katie Meehan and Kendra Strauss, both of whom have provoked ideas and with whom I have developed enduring friendships. Traveling in South Africa with Fiona Lumsden and me, Thulani Ncwane fostered our understandings of the postapartheid landscape as well as nurturing a sense of the possibilities for a better world. That Thulani's participation in making this better world has been cut short seems like one of the cruelest and most stupid blows. Elsewhere in Durban, David Hemson, Patrick Bond, Andy and Tim Gibbs, and Mary Galvin cared for us, taught us, and lifted our spirits. Recently, in London, I have been privileged to work with Johan Moyersoen and, more recently, Jim Segers of City Mine(d). Both here and in Toronto, discussions of Gramsci with Mike Ekers, Stefan Kipfer, and Gillian Hart have transformed my understanding and inspired new work.

At Royal Holloway, I work alongside a wonderfully collegial and supportive group of scholars. Felix Driver first suggested that I should write a book; Mustafa Dikeç urged me to make it better; and many others within the department have supported the work at all stages. Lunches with Mike, Ian, Varyl, Adrian, and Vandana have kept me sane in the midst of ugly changes to the university sector. My PhD students, Richard Bater, Ashley Dawkins, Steve Jones, and Fiona Nash, have kept me intellectually alive and on my toes, as well as commenting on individual chapters.

Several of the ideas from the book were presented in departmental seminars at the University of Toronto, King's College London, and the University of Birmingham, and I am grateful for all the comments. Matthew Gandy suggested approaching the University of Minnesota Press with the manuscript and gave me the confidence that perhaps someone might read it. Jason Weidemann responded warmly and positively, carrying the project through with encouragement and care. Nik Heynen and another anonymous reviewer were kind enough to read the manuscript and comment on it in incredibly helpful ways.

In the past few years, Mike Ekers has become my closest and most supportive academic critic. We have collaborated on several projects, all of which have shaped this book in different ways, and I am deeply indebted to him for many of the ideas. Well beyond what might be asked of even the dearest friend, Mike read and provided detailed comments on a full draft of the manuscript in the blurry nights soon after the birth of his daughter Ella-Jane.

My family retained a bizarre faith in my ability to complete this project, and this has been a continual source of encouragement when things have seemed unmanageable. And none of this would have been possible without Fiona. If our everyday life is structured more than ever by the tyranny of clock time and the fragmentation of social existence, it is also traversed by joy, wonderment, and shared creative journeys. Throughout these creative journeys, Fiona has been my fiercest critic and most loving and beloved comrade. Without her, this book simply would not exist. Together with little Rosa and Lucian, the world seems a more hopeful place.

Introduction: Emerging Moments in an Urban Political Ecology

THIS BOOK IS ABOUT REMAKING OUR WORLD. If this is an overambitious task that smacks of the worst combinations of utopianism and hubris, this is not intended. Indeed the desire for the world to be radically different is, I would argue, a commonplace one: nearly always more a stifled anger than a revolutionary cry, the challenge, surely, is to understand the movement of this anger, to learn from it, build on it, and transcend it in both humble and democratic ways. Because of this, the book develops an immanent critique of everyday life. In no way is this critique built on a disdain for everyday life: on the contrary, some of the most vibrant political ideas have developed when thinkers immerse themselves in the quotidian tumult of their worlds. Marx's thought was transformed by his experience of German communists living, working, and organizing in Paris. Lefebvre's metaphilosophy is shaped by his life with French peasants and resistance activists, and as a taxi driver touring the rougher edges of Parisian life. And Gramsci's *Prison Notebooks* serves as one of the most profound examples of a radical political mind wrestling with the reversal of a revolutionary moment in and through everyday worldviews. None of these thinkers looks down from above with a patronizing sneer for what they see. Rather, they revel in the ambiguities and the latent radicalism that is hidden by the snobbery of their conservative contemporaries.

If remaking the world really is a commonplace desire, the task ahead is also one that requires working with day-to-day reality as it really is. This is far from just a human task. Reality—as is the "human"—is woven out of numerous entanglements of so-called social and natural relations: this implies some kind of an environmental politics and requires some understanding of environmental knowledge production. To me, this must surely respect these entanglements while also struggling to dismantle the false boundaries out of which the social and the natural are separated. Indeed, such tenuous and yet politically toxic boundaries do collapse in the most mundane places; as we work, eat, breathe, play, speak, touch, feel, or hear. The making of these socionatural worlds is a deeply sensuous process, as human and nonhuman are immersed in inter- and intra-action with other people, technologies, species, and things. This book therefore situates arguments about the socio-natural firmly within the sensuous creation of everyday life. Arundhati Roy speaks beautifully of this: "Another world is not only possible, she is on her way. On a quiet day I can hear her breathing." Within the noise and the dirt, the fumes and the concrete, of the contemporary city, I argue that there are conditions of possibility for sensing this alternative world. Not only in quiet moments of reflection but in shared acts of making the world, people hear, feel, and begin to touch the possibilities for making things differently.

If this can indeed be called an environmental politics, and if it grows from the conditions out of which environmental knowledge is produced, it is clearly one that differs in important ways from most of what goes by the name "environmentalism." Whereas there is so much to be admired in the acts of contemporary environmentalists, much also makes me deeply uncomfortable. All too often an environmental politics serves as a force for constructing those antinomies that are so disabling for a politically progressive project, one that respects what Haraway refers to as "becoming with as a practice of becoming worldly."[1] For my mind, too much of environmental politics smacks of a catastrophism that quickly descends into ugly Malthusian claims and disabling dualisms: here an abstract nature is seen to wreak its revenge on an overconsuming-all-too-big human populace. Where I do find inspiration is from environmental-justice activists, for whom the environment is something *lived* as a simultaneously bodily and global process. The environment is as much the toxins circulating through the bodies of people of color as it is a better, more just world to be struggled for. In the quotidian efforts of shack-dwellers seeking to secure access to water, we see a political struggle that mediates such socio-natural relationships. As I will argue, consciousness of both the barriers and the hopeful avenues leading to a more democratic distribution of water has become crucial to contemporary politics in a country like South Africa. Respecting these struggles and extracting the political lessons from them is one of the central aims of this book. It is, nevertheless, an aim that takes me on several, more fanciful, theoretical journeys.

Central to these journeys is the methodological project embarked upon by Marx. For Marx, writing in the nineteenth century, the world was characterized by enormous upheavals: a revolutionizing system of accumulation was utterly transforming Western European societies. Grasping the flux and flow of these monumental shifts, as well as the potentials that lay within them, required a theoretical framework that gave due weight to processes of change. Here he built critically on the giant of German idealism, Hegel.[2] Transforming Hegel's idealist method, Marx developed a dialectical understanding of the world in which the relations between human and nonhuman are of central importance. In a much-cited postface to *Capital*, Marx writes of how he avowed himself "the pupil of that mighty thinker" and "even coquetted with the mode of expression peculiar to him." Nevertheless, "the mystification which the dialectic suffers in Hegel's hands by no means prevents him from being the first to present its general forms of motion in a comprehensive and conscious manner. With him it is standing on its head. It must be inverted, in order to discover the rational kernel within the mystical shell."[3] Thus, Marx dwelt on the practical mediations that shape the world in distinctive ways. This meant privileging relationships and mediations over the form in which they appear (while also recognizing the importance of those forms for immediate reality). At the heart of each of the chapters that follows is an effort to understand the different ways in which subsequent theorists have built on this dialectical method as a way of understanding and changing the different worlds in which they themselves lived.

As many authors have noted, such dialecticism is often captured effectively within works of art. Seeking to reconstruct the existing world within the artistic object is, for many artists, a process of dismantling and rebuilding specific relationships through their own particular eye, ear, or mind. If that object seeks universal appeal (and this is frequently not the case), this is so often because of the way in which it captures what appears latent and unnoticed in the world.[4] All too often, however, the work of art is enjoyed by a privileged few: it thereby reinstates a distance between observer and observed and between the viewer and the object of art.[5] A subsequent aim of this book is to learn from modes of artistic working, but to see this practice as extending deep into the fabric of the modern city. Here I take inspiration from recent urban interventions or critical spatial practices.[6] Within these practices, rather than seeing the canvas or oils as the starting point for a work of art, the city itself is the means of artistic production. Building on Lefebvre's call to make everyday life a work of art and his understanding of the city as a creative product or an oeuvre,[7] I argue that artistic practice can be extended to the socio-natural entanglements of everyday life as part of a progressive political project.

The book is therefore one that seeks to develop an abstract argument through two concrete movements. The first is the postapartheid political project of ensuring access to services. The second surrounds the efforts of artists and activists to transform our cities into laboratories for radical experimentation. I do not seek to cobble these historically and geographically situated practices together, but seek to learn from them both and, from this, to inform the philosophy of praxis that guides this book.

Everyday Life and a Philosophy of Praxis

One of the more radical political moves within Marx's writings, and something that unites the main theorists in this text, is a commitment to a philosophy of praxis. What this means is that progressive ideas are not to be developed and transmitted by hallowed philosophers, political strategists, or religious leaders. World-changing ideas emerge from everyday men and women, whose practical acts make the world as it is. While less explicit in his later writings, this approach is found throughout Marx's work: indeed, *Capital* is best understood as a critique of political economy (the subtitle of the work) from the standpoint of working people. The crucial moment of immanent critique in the book, when Marx descends from the noisy sphere of exchange to the hidden abode of production, relies on the same kind of methodological move as that found in his earlier writings.[8] This philosophy of praxis is intimately related to the transformation of Hegel already referred to.

For Hegel, the movement of the world is shaped by Mind becoming conscious of the alienations that separate it from the natural world. Marx builds on this movement while seeing alienation as a far more concrete condition emerging under specific historical and geographical conditions. Separating the vast majority of people from both their means of existence and the product of their labor is one of the taken-for-granted principles on which the society in which I write rests. For labor to be transformed into a commodity—*the* defining feature of a capitalist society—it is necessary for people to be stripped of the vast majority of resources on which life itself depends, only for these to be packaged, commoditized, and sold back to them. The result is that most people go to work on a daily basis in order to acquire sufficient money to be able to buy those means of existence through a highly complex set of relationships that is defined by the making of things—both more obvious commodities like cars, radios, and televisions, as well as less obvious commodities like people, emotional support, and knowledge—to which we generally do not have direct access ourselves. These commodities are produced for others. The provision of services, from child care to electricity, is shaped by the fragmentation of social relationships and their mediation through monetary exchange. The fragmentation of the work process and the separation of producer from product are central features of the phenomenon of alienation that Marx considers defining of capitalist societies. It is also, as I will argue later, crucial to the stifled anger that lies beneath the apparent stability and immutability of everyday life.

If alienation is produced from the very real conditions in which we work, it is also to be overcome within this process. For it is in the concrete acts of creating their worlds that Marx senses the possibility that subjects might gain consciousness of their role in making histories and geographies and for the world to be made differently. Thus, alienation is not to be overcome by the mind alone. Nor are the antinomies of philosophical thought to be solved in the mind of the philosopher; they are to be overcome through the concrete acts of everyday people as they gain consciousness of their own subjectivity in the world.

If affective labor has recently been emphasized within marxist thinking, this has certain resonances, and also key differences, with earlier debates emerging in the global South and within marxist-feminist thought. In particular, work by scholars such as Harold Wolpe laid great emphasis on the articulation of modes of production and the interconnections between production and social reproduction.[9] For Wolpe, the development of capitalism in South Africa was built on the back of low wages, made possible through the dependence of a wage-labor force on the noncapitalist forms of reproduction taking place in the Reserves. As the conditions on which this social reproduction depended came to be undermined, so wage demands rose. Apartheid emerged in response to such demands and as a means of sustaining South Africa's distinctive form of capitalism. Although feminist theorists have shown the danger in seeing too rigid a distinction between a "sphere of production" and a "sphere of reproduction," Wolpe's argument situates the development of capitalism within the historically and geographically specific practices through which people struggle to sustain their daily lives. In this respect, it has important resonances with the immanent critique of everyday life I develop here.

I use the terms "everydayness" and "the everyday" somewhat liberally and I realize this is not going to hold much sway with some people. Some will claim that it lacks specificity. "What is not everyday? Surely the most powerful in this world also have everyday lives? Are philosophers not everyday?" I will give much greater specificity to the term as the book progresses, and especially in chapter 5. Nevertheless, it is worth mentioning one or two things at this stage. A critique of everyday life, or a critique that emerges in everyday life, is intimately related to a philosophy of praxis: as several authors have shown, it refers to a particularly rich tradition within marxist thought. Here it refers to a form of critical thinking that is immanent to the practice of making the world. It is a critique that emerges through such practices, and it suggests ways of changing those practices in infinitely more progressive ways. Without essentializing, a critique of everyday life suggests where we might look to for change. It implies a distrust of figures of authority. It necessarily means making connections with others—human and nonhuman—and, as I argue, it involves reveling in the dirt and grime, the anomie and the creativity, of city life, where the majority of the world's population now live their lives. Crucially, this book will argue, both a philosophy of praxis and its allied critique of everyday life mean connecting with an understanding of the production of environments.

There's No Such Thing as Nature

Nature, I will argue, has proven one of the greatest stumbling blocks to radical political change. In part this is a good thing. An anthropocentric hubris is stopped in its tracks by the genetic complexity of the fruit fly or the nodding beauty of a host of daffodils. But at the same time, the concept of nature has served some deeply cruel political agendas. Hierarchies have been ascribed to nature: in the UK, our monarch-in-waiting is invested with power because of a "natural" birthright; "racial" differences have, in the problematic history of the discipline of geography, been ascribed to natural climatic variations or physiological characteristics. Such perverse determinisms have reared their heads once more in the work of a range of popular and respected economists and anthropologists. Thus we find Jared Diamond making the peculiar and yet now widely accepted claim that environmental differences have determined the fate of different societies.[10] While rejecting the notion that geography single-handedly shapes the economic prospects of individual societies, Jeffrey Sachs ascribes the fate of much of the world to individual countries being landlocked, too arid, or in tropical zones that favor killer diseases.[11] Environmental determinism and the attendant naturalization of social relations are once again finding a veneer of academic credibility. As various regressive political projects seek to put forward their vision of the naturalness of a heterosexual, nuclear, married unit or the violent appropriation of land for the natural living space of one group or another, it does not take much to begin to question whether nature itself is at fault for our inability to make the world in better ways. And if some marxists have rejected such natural causes, it is not difficult to see why. Nevertheless, in doing so, the marxism that emerged was shorn from its material base.[12] It lost a central part of the distinctiveness of Marx's method: at the very heart of this practical historical materialism is an attempt to understand the mutual coevolution of nature and society as a changing ensemble, a differentiated unity, or a socio-natural assemblage. Our relation with "nature" is one fundamental moment in Marx's overall approach to society: to neglect this relation is to remove a key weapon for both analysis and critique.

More recently, of course, nature has been enrolled in a radically different political project, with the emergence of a highly effective, and still growing, environmental movement. This movement has for many people put questions about the making, unmaking, and remaking of our world more forcefully than any other in the last fifty or so years. While the publication of Rachel Carson's *Silent Spring* is often taken as the birth of this movement,[13] consciousness of the impact of humanity on nature had motivated many groups, both radical and conservative, before the 1960s. The Romantic Movement was one such manifestation. Here nature was raised to quasi-mythical status, inspiring some of the most beautiful works of poetry, music, painting, and drama. In William Blake's more radical and antinomian view, a new Jerusalem is juxtaposed with the "dark Satanic Mills" of encroaching industrialism. Ted Benton draws on these Romantic influences in his reconstruction of a historical materialism with adequate recognition of natural limits, something intimately related to his own work as a field ecologist.[14] They can also be seen to have left an important mark on the work of Marx himself, whose radicalism was fed from the influences of the poets Georg Herwegh and Heinrich Heine.

Nevertheless, the Romantic poets reveled in the image of the isolated individual among the grandeur of nature, something that evoked both humility and a rejection of the vulgar relation with nature being established by the emerging capitalist society. Carson's work, in contrast, had a much broader and more direct appeal, stimulating both popular consciousness of the impact of human actions and, importantly, a reaction on the part of the political class. Active citizens had clear evidence of the detrimental effect of human society on the natural conditions upon which it depended. Ten years later, with the publication of the "blue marble" image of the earth, taken from onboard Apollo 17, this sense of the frailty of nature and of humanity's potential to cause havoc to a unified planet consolidated a modern environmental movement that found its expression in events such as Earth Day and through numerous emerging nongovernmental organizations and more grassroots social movements. In the last twenty years, this modern environmental movement has been increasingly motivated by the growing scientific evidence that human society is in the process of radically altering the balance of gases in the atmosphere and, in so doing, contributing to a significant increase in global temperatures.

Global Environmental Change is now easily the most important rallying point for contemporary environmental politics. It is a major factor for stimulating interest in politics from schools to university campuses, linking a youth politics and a pensioner politics in a way that is rarely seen. In turn, the impact of environmental groups, in alliance with leading scientists, on both domestic and international political agendas has been enormous. While it would be wrong to try to capture the breadth and diversity of contemporary environmentalism, I argue that one result of the growing recognition of global environmental change has, paradoxically, been a resurgence of dualistic understandings of the world. Because of this, nature is increasingly abstracted from the day-to-day reality of people's lives and posited as a force inflicting revenge on the arrogance of human society. As Stephen Daniels and Georgina Endfield write in their introduction to an edited collection in the *Journal of Historical Geography* on narratives of climate change:

> Climate change is presently a Big Story, as both a world-wide chronicle of rising cultural consciousness among political elites and the population at large, as well as the grand, often crisis narratives of environmental change itself, notably those aligned to the graphic rising curves of global warming. Like another major scare story early this century—that of global terrorism—climate change appears millennial in a cultural as well as chronological sense, its moral imperatives assuming an evangelical urgency.[15]

None of this is to deny the very real impact that carbon emissions are having on global temperatures. Rather, my argument in this book is that if we are serious about the ways in which the global environment is being shaped through coevolutionary processes, many of these narratives are decidedly unhelpful.

First, recent narratives of global environmental change can be deeply disempowering. While the sense of a looming crisis has clearly motivated broad swaths of the world population to "act," the apocalyptic accounts of pending disaster have also, ironically, put global futures outside the control of everyday citizens. On the one hand, global environmental change sometimes appears as a train crash that we can see unfolding without being able to influence change in any way. (At other times, the inability of many to "see" this unfolding event merely feeds the doubters. Having had one of the coldest winters in the UK in several years, cynical commentators are reveling in the "myth of global warming.") Apocalyptic visions feed this sense of powerlessness: it matters little how much I reduce my air travel and turn off the light switch, this will not slake the thirst of a burgeoning global population nor alter the recklessness of every other modern consumer. On the other hand, control has been ceded to new technocratic elites and the gadgets they wield. In these figures we invest our hopes for any kind of an alternative. It is they who will build our low-carbon cities, based on assessments of carbon outputs and the abstract quantification of qualities of life. It is they who will redesign our traffic networks, championing the electric car. In the words of one environmental commentator,[16] this has the peculiar effect of inverting the problem: we *should* be asking what kind of world we want to live in, but the question we are asking is how to power this society in low-carbon ways.

This brings me to my second point. Too many narratives of global environmental change depoliticize the processes and relationships out of which climate change is produced. This is what invites the criticism of Erik Swyngedouw,[17] for whom climate-change debates and sustainability discourses represent an emerging postpolitical consensus. In making this argument, Swyngedouw draws on recent attempts to define the truly political. The involvement of high-profile politicians, from Al Gore to Barack Obama, does nothing to diminish the postpolitical status of climate-change debates, motivated as they are by the common enemy, carbon, and the shared fate at the hands of a looming apocalypse. Swyngedouw's swingeing criticism captures the movement of the debate brilliantly. The appearance of democratic debate merely masks a deep political impotence. The unjust relations out of which contemporary ecologies emerge are never questioned within what can be understood as a "police distribution of the sensible." Nevertheless, rather than leading to a rejection of an environmental politics, the direction his critique should take us (and this is what really underlines Swyngedouw's position) is to reformulate such a politics on a ground that *is* characterized by truly democratic debate, for here the possibility of a transformative political moment is once again present. This is exactly the aim of this book. It is an effort to reformulate a politics of the environment in which everyday subjectivity is seen to be at the heart of a revolutionary politics. In order to do this, I suggest we must reformulate environmental politics on the terrain of the quotidian.

The Dialectical Moment

If one response is to rethink the nature of everyday life, this response needs to be grounded in a broader framework that is flexible enough to capture the interaction of what are normally conceived as separate elements: the natural and the social, the historical and the geographical. In doing this, again I draw inspiration from Marx's dialectical approach to understanding the world.

For many, this is the defining feature of Marx's overall method. For others, however, it is an unnecessary source of confusion and obfuscation. Thus, John Roemer equates any discussion of dialectics with obscurantism, writing that "too often, obscurantism protects itself behind a yoga of special terms and privileged logic. The yoga of Marxism is 'dialectics.'"[18] I reject such a position.[19] The dialectical approach that Marx developed is a way of looking at the world that privileges processes and relations over the forms in which they appear. At the same time, it is a mode of abstraction that permits certain relationships to be focused on at certain times through filtering other information. Beyond a way of seeing, Marx's development of the dialectic is also an approach to understanding the world, and it is the manner in which these ideas are presented within Marx's writings.[20]

If Roemer's cynical understanding of the dialectic as obscurantism reflects a broader view,[21] this is also no doubt related to the way in which the Soviet Union went on to develop a science of the dialectic, later codified in Stalin's rigidified dialectical materialism. The perversity of such a rigid framework, given Marx's attempt to capture movement, change, and process, should not go unnoticed. Nevertheless, this demonstrates the ways in which a dialectical approach has come to mean quite different things for different authors in subsequent years. As the book progresses, the contributions of Lefebvre, Gramsci, Lukács, Smith, and Harvey toward a variegated understanding of marxist dialectics should become clearer.

Perhaps, given the range of approaches, David Harvey treads on dangerous territory when he seeks to develop a set of dialectical principles.[22] After noting that Marx never embarked on such an approach himself because this might be self-defeating, he summarizes these principles in eleven distinct theses. Here he draws particular attention to the fluid understandings permitted by such an approach in which the search for processes and relations beyond the apparent permanence of the contemporary moment is prioritized. Writing on this attempt to capture these principles, Nancy Hartsock claims that "Harvey's is one of the best reformulations/distillations of Marx's writing practices I have ever encountered, one that foregrounds the importance of thinking in terms of processes and remembering that every historical form is constituted by its fluid movement."[23] In particular, she praises Harvey for his emphasis on the importance of the *moment*. This is something I will return to on several occasions by looking at the work of subsequent theorists. It is clearly fundamental to our overall understanding of the specificity of the marxist dialectic and to the particular transformation of Hegel in the work of Marx.

In a recent paper, Harvey deepens this methodology, drawing attention to the "deep relevance of a certain footnote in *Capital*" in which both Marx's approach and his particular understanding of "the moment" really come to light.[24] It is worth quoting the central part of this footnote, for it also captures key aspects of the approach I embark upon in this book: "Technology reveals the active relation of man to nature, the direct process of the production of his life, and thereby it also lays bare the process of the production of the social relations of his life, and of the mental conceptions that flow from those relations."[25] Captured in this quote, as Harvey notes, are six conceptual elements or, better still, six moments: technology, nature, relations of production, everyday life, social relations, and ideas. These moments, he argues, should be viewed not causally but dialectically—indeed perhaps the most common error of crude interpreters of Marx is to see one as determining the other, the economy as shaping mental conceptions, for example (this, it should be noted, is a misdirected criticism often leveled at Harvey). Instead, Harvey argues:

> The six elements constitute distinctive moments in the overall process of human evolution understood as a totality. No one moment prevails over the others even as there exists within each moment the possibility for autonomous development (nature independently mutates and evolves, as do ideas, social relations, forms of daily life, etc.). All of these elements co-evolve and are subject to perpetual renewal and transformation as dynamic moments within the totality. But it is not a Hegelian totality in which each moment tightly internalizes all the others. It is more like an ecological totality, what Lefebvre refers to as an "ensemble" or Deleuze as an "assemblage" of moments co-evolving in an open dialectical manner. Uneven development between and among the elements produces contingency in human evolution (in much the same way that unpredictable mutations produce contingency in Darwinian theory).[26]

As coevolving, constituent elements of this socio-natural totality, each moment represents a different level of abstraction (without any hierarchy being implied here) or a different window through which to interpret the world. Hartsock refers to these moments as translucent filters: "First, the filters will determine which features will come to the foreground and which will recede. The filter can be changed as one moves analytically among different moments, and then different aspects of social relations will be revealed. Second, one must pay attention to different historical processes to understand how each moment plays a role in determining others."[27] Throughout what follows, I seek to build on and develop such a method. Nevertheless, if my focus is more on nature and everyday life, it is not because I am privileging these moments above any of the other four referred to in Marx's footnote, nor that I see them as either causal or determinant. Rather, it is because of their relative neglect in other works and their shared importance in understanding the present, the past, and, above all, future possibilities.

Creating Future Possibilities

In recent work, Harvey questions what kind of geographical and anthropological knowledge we might develop that can provide the "conditions of possibility" for an emanicipatory cosmopolitanism.[28] This book seeks to further this project through extending the debate into the realm of everyday life. For Lefebvre, there are two projects within Marx for the transformation of everyday life. One is an ethical project, which involves recognizing true reciprocity between individuals. The second "is aesthetic in nature. It is committed to the notion of art as a higher creative activity, and a radical critique of art as an alienated activity."[29] Indirectly, Lefebvre describes his own project here, but it is fascinating to see how deeply embedded he considers this approach to be within Marx's method. Thus Marx "imagines a society in which everyone would rediscover the spontaneity of natural life and its initial creative drive, and perceive the world through the eyes of an artist, enjoy the sensuous through the eyes of a painter, the ears of a musician and the language of a poet."[30] Although I have difficulties with the romanticization (and essentialization) of "natural life" in the passage, it demonstrates the ways in which Lefebvre viewed his critique of everyday life as, in part, an attempt to extend the boundaries of artistic practice into the quotidian, thereby demonstrating its role in revolutionary change. There is no need to see this as a reversion to some imagined preexisting condition. Nor is there a need to see it as a fixed end point. Rather, Lefebvre places emphasis on the role of creative action in the struggle for a future society.

In different ways, Walter Benjamin ponders the role of the author and the role of the artist in forging revolutionary change in the world: his conclusions approach those of Lefebvre. Creative praxis is clearly crucial to the communist project he envisages, but all too often artists are conceived as having some privileged answer to the deepest problems facing society, thereby reinstating a problematic vanguardism. Avant-garde artistic movements have grown on the premise that somehow this vanguard can both sense the ground ahead and lead society to the realization of the true path to change.[31] In another telling observation, Gramsci comments on such an assumption, stating that "the artistic relationship brings out, especially in relation to the philosophy of praxis, the fatuous naivety of the parrots who think that with a few brief and stereotyped formulas they possess the keys to open all doors (those keys are actually called 'picklocks')."[32] With the role of the artist as his starting point, Benjamin is drawn to the radical transformation of the theater pioneered by Bertolt Brecht in his development of the Epic form.[33] Rather than changing the content of the play, Brecht sought to change the means of artistic production. The alienation techniques that Brecht employed, the introduction of the smoking theater, and his efforts to bring everyday life onto the stage through substituting professional actors for passersby, all contributed to a radically different form of theater in which artistic practice broke free from being confined to a sphere of narrow interest to a part and parcel of the struggle for a new civilization (to take a further term from Gramsci).

Building on the work of Lefebvre in particular, scholarly attention has turned in recent years to the ways in which critical spatial practices reshape an extended artistic field. Thus, Jane Rendell links such work to a critique of public art as wallpaper and a renewed engagement with everyday life.[34] She then details examples of specific critical spatial practices from the subversive work of PLATFORM to the more formal work of architectural practices. As Malcolm Miles notes, critical spatial practices are best characterized as forms of urban intervention that seek to provoke while also generating participatory forms of working.[35] What we see in this expanded field is the ability of insurgent activists to transform the fabric of the city into a means of artistic production. In exactly the way in which Lefebvre called for praxis to expand into the everyday, so, by using the environment of the city as a laboratory for artistic experimentation, we see how urban interventions open up new conditions of possibility for alternative urban futures. Given what has been written earlier about ideas being one moment in a coevolving totality, I would also argue that we need to conceive such practices as inherently socio-natural. They are reducible neither to the social nor the natural but, rather, are part of a continually changing urban assemblage. This book thereby seeks to better understand the limitations and the potentials of interventionist practices in reshaping socioecological totalities in radically democratic ways. It is, in short, and to return to my starting point, a book about remaking our world.

Summary of the Book

The book begins with an attempt to get to grips with the ways in which urban environments are produced. Here I argue that they are best seen as assemblages of social and natural relationships, thereby building on a growing body of work that has been labeled "urban political ecology." At the heart of the work of most political ecologists is an effort to show that the interaction of human and nonhuman serves to create environments in historically and geographically specific ways. Through better understandings of the creation of these environments, an analysis of the power relations structuring environmental processes emerges. In providing both an analysis and a critique—as well as, on some occasions, a normative claim[36]—of existing ecologies, political ecologists seek to overturn depoliticized understandings of environmental processes. Applying this work to the city necessitates an assault on the widely held claim that the city is the antithesis or the enemy of nature. Instead, urban political ecologists seek to demonstrate the myriad processes and relationships through which the city is constructed as a "socio-natural" assemblage.[37] While much of this emerges in the form of critique, it is also built on the premise that any viable environmental future must be constructed on the grounds of an increasingly urban population. The key environmental questions shift from ways in which nature might be preserved in a specific form to the ways in which cities might be constructed based on principles of social and environmental justice.

For several urban political ecologists, one of the richest ways into understanding how urban ecologies are constituted in *this* historical moment comes from Neil Smith's work on the "production of nature."[38] Smith targets his critique at a certain Enlightenment perspective in which nature is treated as an externality or as a resource, something divorced from society but to be appropriated in a variety of different ways. Such an approach makes little sense, he argues, given that nowhere does this pristine nature exist. Instead, the theoretical challenge is to understand the different relationships, forged through concrete activity, that produce nature in specific ways. In the present moment, given the centrality of capitalism as a system of accumulation, this means understanding the way in which nature is produced out of capitalist relationships. Contemplating this, Smith provides a unique interpretation of historical materialism and the grounds for thinking about what he terms a "genuinely human geography." I dwell on Smith's work at length in this chapter because for me, along with several other urban political ecologists, it provides one of the most fecund starting points for rethinking the politics of the city. Nevertheless, Smith's work raises several further problems. In particular, the agential approach he develops appears to give little sway to nonhuman agency. This has been one of the major criticisms to emerge in recent years, building on both theoretical antihumanism and more recent posthumanist critiques. Both of these critiques serve to decenter the conscious subject and disrupt the centrality of human labor in historical materialism. These criticisms are important to new environmental perspectives, and wrestling with such a critique is an ongoing refrain throughout this book. Each of the subsequent theorists, all of whom can be considered to be "humanists" of some form or other, I interrogate along these lines. Nevertheless, in the case of Smith, his approach is better situated within the broader dialectical understanding Smith himself develops. For here, human labor is understood less in a determinist manner and more as one particular moment in the creation of a socio-natural totality. I conclude the chapter with an attempt to think through posthumanist criticisms and to take them forward as a springboard for further chapters.

One of the central contributions of Henri Lefebvre (whose work I turn to in chapter 5) is to emphasize the importance of cultural praxis to Marx's overall vision of a future society. In chapter 2, I begin to excavate aspects of this aesthetic critique through demonstrating the importance of the sensuous to Marx's overall critique of capitalism, his vision of communism, and, crucially, his concept of nature. Although touched on in recent accounts of Marx's concept of nature, the role of the senses is generally neglected. Nevertheless, deeply influenced by both Epicurus and Feuerbach, Marx posits sensuous human labor at the center of both his epistemological and ontological framework. This position is crucial to the practical materialism he goes on to develop. Thus, reality is understood as "sensuous human activity, practice."[39] I begin the chapter by demonstrating the ways in which sensuous human activity is crucial to the environment of one particular postapartheid informal settlement in South Africa. The laboring acts of women serve to create an ecosystem on which the life of this settlement depends. Riven with injustices, this environment is exposed to the whims of commercial relationships. One particular moment of crisis invokes a fiercely politicized response on the part of local women and I seek to question how this can be understood to emerge from the women's situated understandings of the political ecology of the settlement. These understandings, in turn, are related to their sensuous laboring acts.

In placing emphasis on both cultural praxis and the role of sensuousness, a further aspect of the production of nature comes to light. Here the creative element, intimately related to Marx's aesthetic critique, can be emphasized in the remaking of urban political ecologies. Within this book, this permits a closer dialogue with urban interventionist practice. Thus, toward the end of chapter 2, I focus on the work of artists who have sought to traverse the affective domain, thereby showing the manner in which the urban is made up of both material processes and the emotional fabric of daily life. In particular, I look at the work of Christian Nold, whose BioMapping has developed a visual representation of the different emotions that make up the city.

As I have noted, one of the key conceptual moves made by Marx is to situate his critique of capitalism from the standpoint of the worker involved in the act of production. Moving from the noisy realm of exchange to the hidden abode of production permits situated knowledges of the relationships that make reality. Few have taken this conceptual move as far as Georg Lukács, who, in *History and Class Consciousness*, constructs both a critique of philosophy and of existing society, as well as a vision of the revolutionary potentials of the working class on the back of this standpoint theory. In chapter 3, I look at the implications of Lukács's work for an immanent critique of the nature of everyday life. If we understand that urban environments are produced through practical, sensuous activity, Lukács's standpoint theory would seem to have a profound relevance for the emergence of a radical ecological consciousness. In short, those involved in making urban environments through their day-to-day interactions with one another are well positioned for grasping the mutability of the current condition and for generating a transformative politics. Nevertheless, one must encounter several key conceptual difficulties. First, an essentialist understanding of a coherent and unified subject is both oppressive and disabling for a future politics. Instead, in Donna Haraway's figure of the cyborg, one finds a fractured figure that is shaped through historically and geographically specific interactions with human and nonhuman others. Here I seek dialogue with both Haraway's writings and other feminist standpoint theorists. Second, such a critique needs to be repositioned within a critique of everyday life. Here the tainted concept of imputed consciousness in Lukács's work needs to be stripped of its Stalinist connotations. Finally, the socio-natural politics incipient in *History and Class Consciousness* needs to be recovered for Lukács's thesis to work as an immanent critique of the nature of everyday life.

Chapter 4 differs in that it works far more directly with the conjunctural politics of postapartheid South Africa. Seeking to build on the example of Antonio Gramsci, for whom concepts needed to be wrestled with in the concrete and complex realities of the world, I seek to develop a clearer understanding of what happens to this emerging consciousness when it articulates with previous memories of struggle, comradeship, alliances, and enmities. Again, Gramsci sought to build on Marx's philosophy of praxis through respecting the struggles and conceptions of what he was to term "subalterns." Far from romanticizing subaltern conceptions of the world, Gramsci sought to understand how these often fragmented and "incoherent" understandings contained within them a potential core of radicalism that could become a "coherent" view that fused theory and practice into world-changing praxis.[40] In this chapter, I demonstrate how Gramsci might have understood such a transformation as operating within a socio-natural context. On several occasions,

Gramsci refuses to separate nature and society into distinct realms and instead sees the human as being relationally defined through interactions with one another and with the nonhuman world. Through varying combinations of consent and coercion, particular moral and cultural norms come to be stabilized in historically and geographically specific contexts. Once more, this problematic—something Gramsci termed the "operation of hegemony"—can be brought within a dialectical framework that recognizes the internal relations between socio-natures, social relations, relations of production, ideas, technologies, and ideas. Working through these understandings in chapter 4, I look at the contested waterscape of Inanda, a postapartheid informal settlement in South Africa. Here I consider not only how the environment is shaped by particular mental conceptions, but how this then serves to stabilize particular worldviews at particular times. In large part this is related to how ideas articulate with prior historical memory. Building on an immanent critique of the nature of everyday life necessitates a much clearer understanding of this articulation of past and present.

In chapter 5, I take forward the understanding of cultural praxis already suggested in the work of Henri Lefebvre. Although Lefebvre's prolific writing is now well known within the Anglophone world, he is remembered primarily for his theorization of the production of space and secondarily for his critique of everyday life. To date, there has been little engagement with Lefebvre's theorization of nature. In this chapter, I begin to rectify this. Nevertheless, Lefebvre's explicit writings on nature remain somewhat disappointing. He advances an antinomian view in which society increasingly encroaches on nature. This view fails to capture the myriad mixings of society and nature that have forever constituted the nature of everyday life. Nature appears innocent but prone to corruption from society. Such a view is particularly odd given Lefebvre's groundbreaking work on the production of space. Again, space and nature are viewed through dualistic lenses with the former considered a social product and the latter the antithesis of the social. Instead of building on these writings, I argue that it would be more productive to extend Lefebvre's model of cultural praxis to an understanding of nature. Lefebvre saw hope not in extending praxis to the autonomous work of art but in seeking to dissolve the boundaries between praxis and everyday life. John Roberts captures this well: "The outcome is a theory of culture that places a primary emphasis on the extension of the form of the artwork and aesthetic experience into the environmental and architectural."[41] In spite of this, Lefebvre works with an impoverished view of the environmental. If we challenge the antinomian roots to his conceptualization of nature, it becomes possible to extend this cultural praxis to the socio-natural. At the heart of this could be an environmental project that is firmly rooted within everyday life.

In the conclusion, I seek to draw these insights together: it calls for an understanding of the urban that foregrounds a radical socio-natural praxis rooted in the everyday. I link this understanding to a critique of existing environmental practice and to the possibilities within the ruined shards of the latest financial crisis. The task ahead is clearly a monumental one. But it is one that is in the hands of the many and not of the few: because of this, it feels, oddly, more achievable.

Chapter 1 The Urbanization of Nature
Neil Smith and Posthumanist Controversies

ONE OF THE CENTRAL PREMISES OF THIS BOOK is that any framework that seeks to separate nature and society into discrete realms is utterly disabling for a radical and liberatory politics. For an environmental politics, such a view fails practically in that it cannot capture the myriad ways in which the nature we experience on a daily basis is actively constituted through *non*natural processes. The innocent romance of the Lake District, to take just one example, has been shaped by centuries of socio-natural struggle producing complex field systems and grazing patterns that are so fundamental to the natural landscape. For Don Mitchell, writing on the California landscape, this is "the lie of the land,"[1] a lie that eviscerates the laboring acts that produce nature in historically and geographically specific ways. Based on this lie, the target of an environmental politics is misjudged: efforts are thrown behind the attempt to restore the pristine foundations to nature. The radical possibilities for "becoming with" companion species (as Haraway refers to it) are dramatically foreclosed.[2] Furthermore, in disabling liberatory impulses, dualistic understandings give succor to the crudest stereotypes and hierarchies: race, nation, gender, and power become authentic expressions of what nature intended. Unalterable and even desirable, politics must simply accommodate itself to such forces in rational, objective ways. Nevertheless, in spite of the problems that such dualisms generate, much of environmental politics, whether radical or mainstream, argues steadfastly for the defense of an abstract nature, cleansed of any societal influence. Indeed, in a curious paradox, the galvanizing calls for change that have been so genuinely inspiring about the environmental movement in recent years have given cover to the most conservative and repressive impulses elsewhere. Calls for social and environmental justice seem to blur with "optimum populations" and lifeboat ethics.

In recent years, coming from several different directions, there has been a much more concerted effort to challenge society–nature dualisms.[3] In the public realm, both enthusiastically hyperbolic and depressingly apocalyptic statements have pointed to an epochal blurring of a nature–society divide. During the 1990s, genetic engineering became one of the key battlegrounds in such debates as activists sought to show either the possibilities for ridding the world of hunger and misery or the dramatic hubris exhibited by those scientists who chose to defy nature. Asserting such an epochal shift, as many have noted, merely presents a dualism in a different way: either humans have finally managed to understand nature in such a way that real scientific advances can be made, or they have defied a preexisting sacrosanct division between the human and the nonhuman. Again, this fails to recognize the manner in which the very meaning of nature has been constructed through historically and geographically situated representations. *Where* we situate a dividing line between nature and society (said to have been blurred by the arrogance of science) is, in short, historically, geographically, and culturally situated. To paraphrase David Livingstone, nature means something different to a sixteenth-century milkmaid, a seventeenth-century astrologer, and a twentieth-century university student.[4] This is the central claim of work that has highlighted nature as a social construction. The argument here is an epistemological one: its focus has been on the representational practices through which nature is commonly understood and interpreted. Human geographers, in particular, are now so at home with the idea that nature cannot be grasped outside these representational practices that it is almost taken for granted.[5] Indeed there is a rich legacy, extending back through Glacken's monumental work, *Traces on the Rhodian Shore*, to the Berkeley School's work on cultural landscapes, and before that in the work of Vidal de la Blache and the Annales School in which the historical and geographical specificity of conceptions of nature have been carefully reconstructed.[6]

In slight contrast, in recent years, theorists have begun to develop a critique of dualistic conceptions that differs from the epistemological arguments of social constructionism and also from the epochal claims of the populist critique. Here a rejection of the nature–society dualism on both ontological *and* epistemological grounds has started to emerge (although as I will argue, this may have roots in both Vidal de la Blache's writings and also in Carl Sauer's approach to cultural landscapes). Generally, although not always, less concerned with representational practices, this builds on a recognition that the stuff of the world is made up of both things and relationships that simply cannot be separated into boxes labeled "nature" and "society." As Harvey suggests, if we try boxing our own lives in this way, the task soon becomes impossible.[7] Ontologically, the society–nature distinction makes no sense as the basis for an understanding of our world. The majority of this work has emerged within what might be loosely termed actor-network theory and non-representational theory.[8] Within writings on cities, urban political ecologists have sought to develop such an approach through an understanding of the historically and geographically specific ways through which nature is urbanized. The intention of such writing has been to disrupt the idea of the city as the antithesis of nature and to focus on the processes through which the city is constituted as a socio-natural assemblage or, in Harvey's words, as a created ecosystem. Such processes are profoundly shaped by power. Here the city is produced as a particular environment that embodies and expresses, produces and reproduces, the very injustices out of which it also is made.

The starting point for many of these urban political ecologists is the provocative claim by Smith that nature is produced.[9] For Smith, dualistic claims surrounding a discrete nature and society are ideological: they provide a distorted view of what is actually a single reality. Indeed, this much he shares with writings on the social construction of nature. As he eloquently describes, epistemologically two contrasting schools of thought have emerged: one supporting an ideology of nature as discrete from society; the other, stemming predominantly from Marx's transformation of Hegel, recognizes the unity of nature and society and focuses instead on the historical and geographic specificities in which this unity has been constituted. This is where Smith's primary theoretical contribution—the contribution taken forward by urban political ecologists—comes in, as he excavates the basis for this ontological claim around the production of nature. He performs this task through periodizing the production of nature. Inspired by this bizarre thesis, one of my central aims is to explore the fecund terrain opened up if we understand nature to be materially produced. It should be clear that Smith's is a very different argument from one that claims nature is a social construction: it does not find the resolution to practical problems simply in our understandings or representations of those problems (although this is clearly important). Smith's opposition to "philosophical argumentation" marks a more forthright concern with the practicalities of everyday struggles.

The radical potentials in this work are many. Until recently, however, outside urban political ecology, they seem to have been surprisingly untapped. Reading Smith's work in relation to both contemporary debates over nature and alongside some of the historical materialist authors that I will consider develops particularly rich threads for democratic socio-environmental change. This chapter begins, therefore, by laying out the key claims of urban political ecologists before looking at how these claims build upon theses adapted and transformed from Smith's work. The chapter then challenges misconceptions surrounding both bodies of work. It does this as a way of moving beyond the muted standoff between ANT, nonrepresentational, posthumanist, and historical materialist approaches within Geography and also as a way of moving beyond the crude juxtaposition of humanist and posthumanist positions. The chapter concludes by considering the ways in which such radical possibilities might be developed through a dialogue with the philosophy of praxis in Marx, Lukács, Gramsci, and Lefebvre. In many ways, the production of nature permits a radical rethinking of such work.

The Urbanization of Nature

Where the starting point should be for a critical analysis of the city has provoked ongoing debate among urban theorists. This has led to a deeper questioning of the entire concept of the urban. David Harvey poses the problem neatly, writing that "the thing we call a 'city' is the outcome of a process we call 'urbanization.' But in examining the relationship between processes and things, there is a prior epistemological and ontological problem of whether we prioritize the process or the thing and whether or not it is possible to separate the process from the thing embodied in it."[10] In a classic statement of his overall approach to urbanization, Harvey goes on to call for an approach that regards processes as more fundamental, while recognizing that all processes are mediated through the things they produce, sustain, and dissolve.[11] Urban geographers should focus not on the city as their primary object of analysis but on the process of urbanization as it produces, sustains, and dissolves individual cities in historically and geographically specific ways. A central aim of work in urban political ecology has been to pick up this approach and to demonstrate that urbanization involves a material transformation of lived environments: because of this, urbanization must be understood as a socio-natural process and never in a one-sided manner. Picking up on Harvey's analytical shift from "the city to urbanization," Swyngedouw and Kaika write that "the environment of the city (both social and physical) is the result of a historical geographical process of the urbanization of nature."[12] Here they not only draw out a process-based understanding of the city but in the process collapse the long-standing dualism between the city and the country. The city as an object of analysis is thereby displaced in favor of a historical geographical materialism that highlights the processes and relationships out of which it is produced and reproduced. They conclude by arguing that "viewing the city as a process of continuous, but contested, socioecological change, which can be understood through the analysis of the circulation of socially and physically metabolized 'nature,' unlocks new arenas for thinking and acting on the city; arenas that are neither local nor global, but weave networks that are almost simultaneously deeply localised and extend their reach over a certain scale, a certain spatial surface."[13]

Swyngedouw and Kaika's call for such work has become a central plank in the theoretical framework that has sometimes been termed "urban political ecology." Although sharing many affinities with the broader field of political ecology, the historical materialist roots are much more evident in the work that is associated with a loosely based (and now diasporic) Oxford school of urban political ecology. Perhaps its clearest expression is found in the edited collection by Heynen, Kaika, and Swyngedouw,[14] which takes as one of its starting points Harvey's often-cited expression that the city is a created ecosystem and that "in a fundamental sense, there is nothing *unnatural* about New York City."[15] For Heynen, Kaika, and Swyngedouw, "questions of socio-environmental sustainability are fundamentally political questions. Political ecology attempts to tease out who (or what) gains from and who pays for, who benefits from and who suffers (and in what ways) from particular processes of metabolic circulatory change."[16] To define urban political ecology as a coherent body of thought (or even as a loosely defined "school") is, perhaps, to oversystematize what is actually a remarkably diverse set of approaches to considering the politicized environment of cities. Influences have been similarly broad, with writers drawing variously on William Cronon's magisterial account of the development of Chicago in *Nature's Metropolis* to environmental-justice debates and a focus on the development of environmental social movements within the city.[17] Nevertheless, there has been a concerted effort to read the city as one *moment* in ongoing socio-natural processes of transformation. Perhaps it is no chance that the focus of the Frankfurt School—on socio-natural metabolisms—has returned, albeit with an intent that is far closer to Smith's than Schmidt's. Empirically, work has ranged across cities in both the global North and South with foci on anything from urban forests to fat in sewerage infrastructure.[18]

Some of the key arguments of urban political ecology have been best captured in work on water. Drawing on actor-network approaches but, above all, on historical materialist accounts of the production of nature, Swyngedouw has sought to demonstrate the manner in which water and social power come to be fundamentally intertwined. As a socio-natural hybrid, water embodies and expresses the relations through which it is reproduced. The flow of water through a city makes life possible and, at the same time, structures the ways in which that life is lived. It produces and reproduces hierarchies, naturalizing those relations that are socially produced and socializing the dependence of the city on the environments out of which it is produced. Key tropes for interpreting the political ecology of cities in Swyngedouw's view are therefore those of circulation and metabolism. As water and capital become fundamentally intertwined through a metabolic process, so circulation becomes a key element in the production, reproduction, and survival of capitalist relations of production.[19]

The twin focus on circulation and metabolism has helped to transform research on water politics and opened up numerous new avenues for radical scholarship. Maria Kaika's research on Athens manages to traverse the material and symbolic, showing how large-scale acts of water infrastructure engineering serve powerful ideological functions.[20] Similarly, Matthew Gandy's rich study of New York City demonstrates how the urbanization of nature not only shapes the material form of the city but also the way in which power circulates and is stabilized.[21] Such work opens up a new agenda for theorizing the role of produced natures within the stabilization of specific social formations and the making of specific subject identities.[22] Within my own work, I have sought to place emphasis on reproductive acts in the provisioning of households with water.[23] This is, in part, a return to some of the core concerns of Smith's work. It also opens up areas of cross-fertilization with historical materialist antecedents to Smith's writing and, more recently, with work in actor-network and nonrepresentational theory. In order to develop this claim, nevertheless, we need a far clearer understanding of the roots of the "production-of-nature" thesis. This requires cutting through several of the misconceptions that surround Smith's seminal argument.

The Production of Nature

On the surface, *Uneven Development* appears to be a book of two disconnected parts. On the one hand, it is a book about the production of space and the uneven development of local, national, and regional economies. Because of the prominence of arguments about the production of space, all too often the text is lazily included within parenthetic references to Lefebvre's *Production of Space*. Here distinctive arguments are elided. While there are indeed interesting connections between Smith's work and Lefebvre's, they differ in fundamental ways. At the heart of these differences are their understandings of nature. For Smith, an ontological claim about the production of nature provides a firm basis for interpreting the production of space. The concrete activity of making space in different ways is something only spectral within Lefebvre's references to spatial practices. In contrast, the production of space within Smith's work can only be interpreted through concrete acts of producing nature: hence a book of apparently two parts—the one about nature, the other about space—even if the two parts are in fact inseparable. In recent years, it is to the former part of Smith's argument that geographers have turned their attention, albeit through once again ceding to Lefebvre the genesis of an understanding of the production of space. However, as I will demonstrate, bringing space and nature back together opens up productive possibilities.

Smith's objections to dualisms are similar to those running throughout this book. Dualistic understandings of nature and society, in which nature is externalized as a resource, make no sense ontologically. And, furthermore, they provide real problems for a progressive politics. In spite of this, as he demonstrates, these fractured conceptions are reproduced within both scientific and literary practices. Here nature is, on the one hand, the external or nonhuman world and, on the other, the world including humans. While many have sought to challenge the reifying tendencies of the latter—through questioning the biological constitution of race, the rigid boundaries that separate nations, or the natural forces that compel competition within capitalist economies—these efforts ultimately founder if the former conception (that of an external nature) remains unquestioned. As Smith writes, "for 'human nature' to fulfil its ideological function there must be a separate nature with its own inviolable powers." And again, "the possibility of the socialization of universal [human] nature is ultimately denied not on the basis of historical experience but by the contradiction with external nature."[24] Kant struggled with this contradiction. Hegel appeared to overcome it, only to dissolve nature into history. However, through Marx's radical transformation of the Hegelian dialectic, we find a methodological basis from which to understand nature in nondualistic terms. Nature, in short, comprises a differentiated *unity* of the human and nonhuman. For Marx, this dialectical unity is achieved through practical activity and, in particular, the act of production. Labor thereby mediates a metabolic process in which human and nonhuman are inseparable and codetermining. To adopt Latour's terminology, it provides a basis for the understanding of socio-natures.

As a prelude to his exploration of the production of nature, Smith accepts that it may indeed be possible to detect within Marx's work something of a dualistic conception of nature. Demonstrating this, he focuses on Alfred Schmidt's *The Concept of Nature in Marx*.[25] As a key "second generation" theorist of the Frankfurt School, Schmidt was lauded by critics in the 1970s and 1980s for providing the firmest scholarly understanding of Marx and nature. Drawing out what he felt to be some of the more Kantian influences on Marx only appeared to add more to Schmidt's rich interpretation. Smith, nevertheless, notes fundamental inconsistencies in the overall approach and, controversially, goes on to claim that Schmidt provides one of the most elaborate examples of "the bourgeois conception of nature" while claiming this to be Marx's. This argument could easily be consigned to history if it did not encapsulate so many of the difficulties of subsequent eco-marxist and eco-socialist thought, which has been plagued with how to overcome such disabling dualisms. Schmidt's thesis ultimately falls for the same reasons as his Frankfurt School mentors.[26] The politics that emerges, as with much of what might be termed Western marxism, is ultimately paralyzing rather than liberating, owing to the development of a claim about the enduring struggle between man and nature. As Martin Jay has shown, the Frankfurt School's focus on nature marked its final march away from orthodox marxism. Instead of the class struggle, the focus became "the larger conflict between man and nature both without and within, a conflict whose origins went back to before capitalism and whose continuation, indeed intensification, appeared likely after capitalism would end."[27] Change seems futile within such a framework: indeed, claiming that Marx was a utopian for daring to imagine alternative relations through which labor might mediate socio-natural processes, Schmidt then suggests that this utopianism cannot overcome a fundamental drive within humanity to dominate nature. In short, Schmidt claims, there is a distinction within Marx between his claims over nature and his economic claims. Marx-the-utopian assumes that the domination of nature will cease with the transition to a fundamentally different mode of production. Thus, Schmidt writes: "The new society is to benefit man alone, and there is no doubt that this is to be at the expense of external nature.... [E]ven in a truly human world there is no full reconciliation of Subject and Object"[28] Although doing so much to dismantle dualistic claims, Schmidt falls back on a conception of external nature and projects this back onto Marx, foreclosing virtually all the radical possibilities within Marx's transformation of the question of nature.

In method, even if not in spirit, this echoes other common socialist ecological positions. For example, as I explore later, the domination of nature is a thesis around which Lefebvre appears to drape a concern for the loss of nature while simultaneously naturalizing his conception of space and rendering nature inert, passive, or dead. As in Schmidt's work, eco-socialist developments of the "domination of nature thesis" are based on a particular conception of external nature that opens the way for an ontological divide between the social and the natural. Within this, dominating nature is seen as an unavoidable drive. For several within the Frankfurt School, the concept of nature assumes a teleological role. Society unfolds in such a way that can only result in the eventual domination of internal nature and of human over human. There is quite simply no way out. In an ultimately liberatory rejoinder, Smith seeks to recover what he sees as a more honest reconstruction of the concept of nature in Marx:

> The negative triumphalism of the "domination of nature" idea begins with nature and society as two separate realms and attempts to unite them. In Marx, we see the opposite procedure. He begins with nature as a unity and derives as a simultaneously historical and logical result whatever separation between them exists. . . . Instead of the "domination of nature," therefore, we must consider the much more complex process of the production of nature. Where the "domination of nature" argument implies a dismal, one-dimensional contradiction-free future, the idea of the production of nature implies a historical future that is still to be determined by political events and forces, not technical necessity.[29]

This statement speaks to several of the misconceptions that surround Smith's writing, especially those that criticize the "production of nature" thesis for foundering on its latent "dualistic assumptions." Surely, I argue, such criticisms move in the opposite direction from both the intent and the word of Smith's work. They are, perhaps, more Schmidtean than Smithean. Instead, understanding nature as produced is to make a concerted effort at developing a politics of nature that is both nondualistic and nonanthropocentric, even as it retains a fundamental belief in the ability of men and women to make their own histories and geographies (albeit not under conditions of their own choosing).

If Smith's argument differs in fundamental ways from Schmidt's, it is in part down to the meticulous reconstruction of a jigsaw puzzle comprising pieces from *The Economic and Philosophical Manuscripts, The German Ideology, Grundrisse,* and *Capital.* Emerging from this puzzle is a contextually rich understanding of the importance of history and geography to the metabolic process described by Marx. Linking stages in the development of the production of nature to the periodization Marx offers in the *Grundrisse*, Smith focuses predominantly on "simple production," "production for exchange," and "production on a world scale." Throughout these different modes of production, nature is understood to be a co-constitutive agent, making people through the actions they perform. As Marx writes in the third volume of *Capital*, even as the labor process changes, its simple elements remain the same. Building on such a claim, Smith states directly "in its ability to produce nature, capitalism is not unique. Production in general is the production of nature."[30]

Fundamental here is the dialectical relationship between humans and nature. Nevertheless, these mutually determining relationships change depending on the form of production at any given moment. Thus, if Smith's periodization is one that applies only to Western capitalist societies, the fundamental claim—that nature is produced—would, presumably, be applicable universally. It would quite simply be impossible for humans (and nature, as we have come to understand it) to survive if such a relationship did not exist. As Marx argues, nature is our inorganic body:[31] existence without this metabolic exchange is inconceivable.

If objects are produced for the direct consumption of oneself or one's family, as they are in what Smith (drawing from the *Grundrisse*) terms simple production, the relationship is straightforward and direct. With the growth of a surplus—something necessary for communities to be able to protect themselves from periods of scarcity—and the growing possibilities for goods to be produced for exchange rather than merely for direct consumption, a new type of relationship begins to develop. During this period, commodities emerge. Embodying both the use value typical of periods of simple production and an exchange value, commodities permit the exchange of otherwise entirely different objects because of a commonly agreed quantitative value. Within simple exchange systems, ten deer can be exchanged for five pairs of shoes: these items can then be exchanged for two coats and so on. A universal measure of value greatly facilitates even the simplest of exchanges by reducing the disparate qualities of an individual commodity into a single, universally recognized measure. Money, in some form or other, comes to be this universal and fixed measure of exchange value. With the appearance of money, nature is produced on an extended scale: it includes much beyond the immediacy of one's direct subsistence and a societal nature is thereby produced. At the same time, the work process itself fundamentally transforms. Divisions between manual and mental labor begin to emerge, permitting different forms of consciousness that are fundamentally related to the production of nature in its different forms.[32]

During this period of production for exchange, a distinction between first and second nature emerges. The distinction between first and second nature appears to be introduced as a way of connecting Smith's argument to Marx's transformation of Hegel. Although often posited as key to Smith's thesis,[33] this distinction is actually far more of a distraction: it only serves to add confusion to the overall argument by verging on the reproduction of a dualistic argument. When production takes place for exchange, Smith argues, a distinction between first and second nature emerges. Whereas for Cicero the distinction between first and second nature was one between objects and humanly produced objects, in the eighteenth century the latter came to include institutions as well as legal, economic, and political rules. Smith thereby distinguishes between first and second nature on the basis of *exchange value:* first nature is seen to relate to use value; second nature refers to something possessing both use and exchange value. Latterly, this distinction is taken further in the final period that Smith covers—capitalist production.

The transformation of society that takes place in the transition to a capitalist society is defined by the division between two classes, one in which the majority depend on the sale of their own labor-power in order to survive. Capitalism is thereby built on an antagonistic social relation. Demonstrating this, in its clearest possible fashion, Marx relates the story of unhappy Mr. Peel. Exporting both workers and means of production to the American colonies in the hope of transplanting the capitalist production process, Peel fails to realize that capital is fundamentally a social relation: "Once he arrived at his destination, 'Mr Peel was left without a servant to make his bed or fetch him water from the river.' Unhappy Mr Peel, who provided for everything except the export of relations of production to the Swan River."[34] Surrounded by bountiful means of subsistence, the workers that Peel has shipped to the colonies can of course survive freely without having to work for him. They flee, presumably to live a much happier life elsewhere while producing for one another directly. As this "unhappy" example shows, capitalism's survival depends on the bourgeoisie's ability to continually enforce a radical divorce between workers and their means of existence. Not surprisingly, the production of nature is utterly reconfigured in a society dependent on actively reproducing this relationship. Above all, the exchange relation—key to Smith's distinction between first and second nature—rises to the fore. Thus, first nature is progressively produced from *within* and as part of second nature. Smith writes:

> Under capitalism, then, the role of exchange-value is no longer merely one of accompanying use-value. With the development of capitalism at a world scale and the generalization of the wage-labour relation, the relation with nature is before anything else an exchange-value relation. The use-value of nature remains fundamental, of course, but with the advanced development of productive forces, specific needs can be fulfilled by an increasing range of use-values and specific commodities can be produced from a growing array of raw materials. The transformation to an exchange-value relation is something achieved in practice by capitalism. Capitalist production (and the appropriation of nature) is accomplished not for the fulfilment of needs in general, but for the fulfilment

of one particular need: profit. In search of profit, capital stalks the whole earth. It attaches a price tag to everything it sees and from then on it is this price tag which determines the fate of nature.[35]

It is worth quoting this passage at some length because, again, I think it speaks to some of the misreadings of Smith's argument. Even within the work of an absolutely first-rate theorist such as Noel Castree, we find the bizarre claim that Smith "used Marxian economics to argue that *capitalism* has replaced a nonhuman 'first nature' with a socially produced 'second nature.'"[36] Elsewhere, Castree refers to Kloppenburg, who, he claims, "exemplified empirically what Neil Smith argued theoretically: that capitalist firms increasingly fabricate a non-natural nature with potentially dire human and environmental consequences."[37] If true, such a claim would not only be reductionist in the extreme, but it would also fail to deal with the really important ways in which capitalism has transformed what has *always* been a differentiated unity—in this it reworks a process that has been under way at least since the presence of humans on the earth into one defined by *specific* social relations. Castree's understanding appears to rest on a separation between a "natural nature" prior to capitalism and the "non-natural nature" he refers to that replaces the former through the actions of capitalist firms, GM engineering, and so on.[38] Here we are back to the epochal understanding of the blurring of any nature–society divide and one that is applied to a reading of Smith.[39] In contrast, Smith makes it clear that under capitalism "the distinction is now between a first nature that is concrete and material, the nature of use values in general, and a second nature which is abstract, and derivative of the abstraction from use value that is inherent in exchange value."[40]

These claims have at least two important bearings on any radical politics that might emerge. Again, I think this is lacking in many of the secondary readings of Smith's work. First, there can be no way of returning to some assumed authentic relationship in which nature is purified of human contact (a "natural-nature" devoid of capitalist firms in Castree's terms?). For this would not only be devastating for all humans, it would also potentially disrupt those very many ecosystems dependent on this metabolic process themselves.[41] Here we may think of the many urban brownfield sites in which species diversity has flourished within the most overtly manufactured of landscapes. Environmental problems are not thereby understood as being about ridding nature of humans; instead, such problems must be situated socially and historically. This brings me to the second point: Smith points us toward a relational and historicized ontology in which nature comes to be defined in the present moment by the specific relations through which it is produced. Seeking a kinder and more democratic production of nature thereby depends on transforming capitalist social relations. This is a fundamentally different goal from that of the vast majority of contemporary ecological movements. Moreover, Smith comes to this conclusion through a nuanced development of Marx's thought, without having to tack on a concern for the environment, or to develop the somewhat overstretched claim that Marx was somehow a proto-environmentalist.[42] Smith's piecing together of the jigsaw puzzle in which he presents the "production of nature" is a superb example of a theorist developing Marx's method (while never straying very far from the text, in Smith's case) in order to present a workable theorization of nature based on marxist principles. Surprisingly, perhaps, this work has not really been advanced by many theorists. Again, the radical and liberatory potential still remains unexplored.

Controversies

This is not of course to claim that Smith's work is without fault. Nevertheless, a large proportion of the criticism that is commonly leveled at the production of nature thesis can, I suspect, be put down to both a misreading—Smith all too often becomes Schmidt—and also to the linguistic difficulties most writers, Smith included, encounter in dealing with socio-natures. Smith is acutely aware of these difficulties and finds hope in a dialectical conception of nature and society that captures both unity and difference (and, above all, internal relations). Nevertheless, as with most, he is prone to a discussion of a social relation *with* nature or of capital defining the fate *of* nature. As this book is no doubt a demonstration of the point, it is surprisingly difficult to free ourselves of the linguistic convention of separating nature and society, even if the thesis we develop is an explicit attempt to demolish such a misconception once and for all. However, this should surely not be the deciding factor on which a thesis stands. If we suspend judgment on the linguistic questions, Smith must surely be seen to provide an utterly remarkable and ultimately liberating take on the unity of socio-natures.

Beyond this, however, there would appear to be a more general frustration with the centrality of the human within the theory. Although he does not regard it as disabling, Castree claims that "Smith shares with Schmidt, Harvey and Grundmann a strong sense of anthropomorphism."[43] Indeed, many others would argue that the outcome of any claim around the production of nature is to develop an anthropocentric outlook in which nature is sidelined or rendered passive and inert. The agency of nonhumans is said to be denied by a theory that necessarily depends on the practical activity of humans and humans alone for its understanding of nature. The optimistic calls within Smith's work for "a genuinely social nature" appear to reinforce this perception and seem to conjure up visions of the Soviet Union's draining of the Aral Sea or China's more recent cloud-seeding projects. And Smith's claim that the victory of working-class revolt "would bring with it the historically unique opportunity for human beings to become the willing social subjects not the natural subjects of their own history" appears for some to neglect the myriad connections with nature that make action possible.[44] They undermine the nondualistic intent of Smith's work by seeking to overemphasize the agency of human subjects. Many of these criticisms have come from actor-network approaches and, more recently, from work on nonrepresentational theory. One of the central thrusts of this work has been to demonstrate the ways in which nature is enacted or performed. Far from dead, inert, socially constructed, or produced, the emphasis is placed on the liveliness of natures.

Although a comprehensive critique of Smith's work has not been fully articulated, criticism has been leveled at a range of dialectical approaches of which Smith's is considered a leading example. Thus, for Sarah Whatmore, dialectical reasoning does not "provide a radical enough basis" for theorizing the socio-natural as it tends toward a "binarized" view of the world. She goes on to claim, in a perverse interpretation of Harvey's position, that "far from challenging this *a priori* categorization of the things of the world, dialectics can be seen to raise its binary logic to the level of a contradiction and engine of history."[45] Similarly, within Jamie Lorimer's otherwise fascinating discussion of posthumanism, marxist political ecology is juxtaposed with the vitalist work of Deleuze: the former is dismissed because of a focus on "scalar politics and the disguised binaries of the nature–society dialectic."[46] In the same way, Steve Hinchliffe takes his criticism from Latour to claim that there is "a tendency in dialectics to systematize, to render relations as contradictions and to eventually pose nature and culture as pure ontological categories that no one can reconcile."[47] The problem, it seems, is that Smith, Harvey, Heynen, Kaika, and Swyngedouw all rely on a failed conceptualization of the mixings of socio-natures. In spite of the fact that each one targets a theoretical project at the false ontological categories of nature and society, these authors are criticized for reproducing in a different form some kind of an ontological separation. Suffice to say that this claim of irreconcilable ontological categories, of disguised binaries, and of engines of history driven by contradiction is unrecognizable from the account given above. It is a misreading of the approach and a misreading that has been reproduced without question, neglecting what

Castree regards as remarkable affinities between "the new dialectics" and ANT approaches.[48]

Nevertheless, a deeper divide might reside in the antihumanist and posthumanist foundations to more recent vitalist approaches. In some (often linear) readings, the "production of nature" thesis appears to be consigned to a poisoned humanist past, and its calls for a "genuinely humanising geography" unashamedly place human agency at the forefront of a radically transformative project. However, the situation is actually far more confusing than the manner in which it is sometimes presented. To take Neil Badmington's definition of humanism, on which he constructs his posthumanist alternative:

> Humanism is a discourse which claims that the figure of "Man" [sic] naturally stands at the centre of things; is entirely distinct from animals, machines and other nonhuman entities; is absolutely known and knowable to "himself"; is the origin of meaning and history; and shares with all other human beings a universal essence. Its absolutist assumptions, moreover, mean that anthropocentric discourse relies upon a set of binary oppositions, such as human/inhuman, self/other, natural/cultural, inside/outside, subject/object, us/them, here/there, active/passive, and wild/tame.[49]

Whether or not Smith privileges human agency, Badmington's view of humanism is not one that could be applied to such work. For Smith's is surely a historically and geographically situated understanding of nature, and one of his primary targets is those binaries then listed by Badmington. Such a definition of humanism is both too blunt and based on too much of a straw figure for it to be able to have any real conceptual weight. Who after all would still accept such a crude account of their work? The claim later that it is "principally in the wake of Jacques Derrida" that such binary oppositions have become less certain is a ridiculously partisan shot that is as laughable as it is wrong.

In the work of a figure such as Haraway, nevertheless, we find a far more nuanced articulation of something that resembles a posthumanist position (more recently she has rejected the term "posthuman" in favor of "companion species," although her interest in "how we became posthumanist" remains).[50] Here the target is not the imagined dualisms in a "Marxian" position—Haraway's groundings in Marx are far too rich to accept this uncritically. Rather, it is the prominence given to the category of human labor and the human subject that diminishes the importance of companion species. Thus, writing on Foucault, she argues that "he understood that this is about the provocation of productivities and generativities of life itself, and Marx understood that too. But we've got to give that a new intensity, as the sources of surplus value, crudely put, can't be theorized as human labour power exclusively, though that's got to remain part of what we're trying to figure out. We can't lose track of human labour, but human labour is reconfigured in biotech-capital."[51]

In *When Species Meet*, Haraway furthers this posthumanist claim through an understanding of "encounter value" as part of her effort to write *Biocapital Volume 1*.[52] Interestingly for Smith, it is "the centrality of labour" in Haraway's "revisioning of biotechnological nature" in which the theoretical power of her approach lies.[53] Sensuous laboring activity is absolutely crucial to Smith's overall formulation: Haraway's questioning of the determinate role of human labor points to something of a conceptual dilemma for any environmental politics that might emerge from this. Given this, we need to establish the extent to which human labor *is* the determining factor in the production of nature, for any claim to an understanding of nature as a differentiated unity would seem to falter on such grounds. Indeed, in what has become a classic exchange, it would appear to be on exactly these grounds that Benton rejects Grundmann's interpretation of marxism as an adequate basis (in unreconstructed form) for an ecological politics.[54] And Benton's unwillingness to engage with notions of the production of nature would seem to lie in his reaction to the human exceptionalism that seems to open up a dangerous privileging of human over nonhuman. Indeed, Benton's engagement with natural field history suggests a similar form of "encounter value" to that of Haraway (whose more recent encounters have been with companion species in the form of dogs) and leads to a similar rejection of notions of production.

Toward a Theory of Moments in the Production of Nature

A way out of this apparent fix is provided, I would argue, not in rejecting Smith's approach but in deepening the dialectical foundations, and especially the philosophy of internal relations, on which it is based. This is something I will seek to do in subsequent chapters by bringing the "production of nature" thesis into conversation with a range of historical materialist approaches to theorizing the everyday. Nevertheless, before doing that, I think we might try to push the theory forward on the grounds on which it was developed; above all on a nondualistic understanding of the socio-natural that is rooted in Marx's overall dialectical approach. As I discussed in the introduction, this dialectical approach is never stated explicitly within Marx's writings and because of this it has been the source of much heated controversy since, ranging from a Stalinist application of dialectical materialism to the nondialectical dialectics of Derrida. Smith's understanding of the dialectic would seem neither teleological nor determinist; rather it serves to emphasize the mutual coevolution of human and nonhuman. Labor is one moment in this dialectical process and it may be seen as a crucial point of leverage for a future transformative politics, but I am not convinced it need be seen as the *only* point of leverage.

Although Smith has not sought to expand on this philosophy of internal relations in a comprehensive manner, it has become one of the guiding missions of David Harvey, who, in more recent works, has integrated Smith's work on the production of nature as a fundamental component to his work. In one of his more recent attempts to distill these dialectical principles, Harvey finds suggestions in a footnote in *Capital*.[55] Here Marx suggests that an understanding of the socio-natural totality is best expressed through six moments that both embody and express change in each of the other six moments. These are the relations of production, social relations, mental conceptions, technology, relations with nature, and everyday life. None of these moments should be seen as determinant in any strict sense of the term, and Harvey spends some time rejecting determinist understandings, stating that many of the problems with specific variants of marxism (and countless other bodies of social theory) emerge from the privileging of one or other of these moments over another. He cites environmental determinism, class determinism, idealism, and technological determinism as key examples that have failed to keep these distinct but internally related moments in tension. In contrast, Harvey claims that "the bounded interplay between these six socio-ecological moments" is constitutive of place and nature. He makes the richly suggestive claim that "when we look back at some of the best forms of historical regional geography, such as that produced by Vidal de la Blache, what we see is a way of understanding the production of regionality through a coevolution, over space and time, of the moments that Marx defines."[56]

Perhaps here we have some evidence that the position that Smith excavated need not center on the determinant role of human labor. Instead, human labor should be viewed as one of several constitutive moments in a process of mutual coevolution. Nevertheless, Harvey goes on to confuse things by appearing to bring us back to where we started when he asks: "How are we to reconcile this way of thinking with the proposition that the labor process lies at the core of the dialectical relation to nature?"[57] Now we are straight back at the problem Haraway recognized. Harvey's response is not at all clear, and he moves between statements demonstrating Marx's determinism and reductionism and his own claim that at the heart of Marx's method is a nondeterminist approach. He concludes by holding firm to the "method of moments" while claiming that the physical transformation of nature through human labor is the only object point in this coevolutionary process where we can physically "measure" impacts. It is worth quoting him at length:

> The method of moments earlier outlined is in no way violated under this conception. The material measure of the relation to nature does provide a solid baseline for judgement, and this seems to introduce a certain asymmetry into how relations among the moments unfold. We cannot, in short, eat and drink ideas, and our material reproduction as species beings within nature has to recognise that elemental fact, even as we freely acknowledge that our species being is about far more than just eating and drinking. But at the heart of all these dialectical interactions among the different moments, at the core of the process of coevolution, lies the foundational question of the organization of human labor because it is through the material activities of labouring that the crucial relation to nature unfolds. Any project that does not confront the question of who has the power to organize human labor and to what purposes and why is missing the central point. Not to address that point it to condemn ourselves to a peripheral politics that merely seeks to regulate our relations to nature in a way that does not interfere with current practices of capital accumulation on a global scale.[58]

So, while never determinant, laboring, material activity comes to acquire a particular importance owing to its absolute centrality to human survival and, crucially, because of the political leverage that it opens up.[59] To me, this provides some kind of a compromise and a way of acknowledging the weight of posthumanist critique while also refusing to cede what is a crucially important aspect of political mobilization in late capitalism. Perhaps the compromise seems unsatisfactory, and I will return to it on several occasions. In the process, I will argue for the importance of broadening our understanding of material activity to encompass reproductive and creative work. This to me seems a further aspect to the production of nature that is often neglected.

Conclusions

Both on epistemological and ontological grounds, dualistic understandings of nature and society should be rejected. They harbor deep conservatisms and fail to capture the way in which life is made through defying such mythical boundaries. Stating this is the easy bit. Figuring out what it means for an environmental politics, an understanding of the urban, and for a program of political change is a far more complex problem. In this chapter, I have sought to build on one particular body of theoretical work in order to begin to untangle this. Using Smith's work, I have argued that practical activity is crucial to the mediation of socio-natural processes. Nature is therefore that which is produced, not the pristine sphere, untainted by human activity, that we sometimes imagine it to be. Situating practical activity both historically and geographically means attending to the relations through which it is structured. In the contemporary moment, this requires a focus on capitalism as a system of accumulation. Nature under capitalism comes to embody both a use value and an exchange value. The dominance of the exchange-value relation serves to demarcate a specific moment in the production of nature. The reduction of the myriad qualities that make environments so distinct to a set of abstract quantitative inputs and outputs is one feature of the contemporary moment. The environmental politics that I seek to develop in the remainder of this book is one that might start from such principles, while also recognizing the possibilities for new ways of thinking, new worldviews, and an immanent critique of the nature of everyday life.

Chapter 2 Sensuous Socio-Natures
The Concept of Nature in Marx

> This communism, as fully developed naturalism, equals humanism, and as fully developed humanism equals naturalism. . . . Society is therefore the perfected unity in essence of man with nature, the true resurrection of nature, the realized naturalism of man and the realized humanism of nature.
> —Karl Marx, *Early Writings*

> Of all philosophers, Marx understood relational sensuousness, and he thought deeply about the metabolism between human beings and the rest of the world enacted in living labor . . . however, he was finally unable to escape from the humanist teleology of that labor—the making of man himself. In the end, no companion species, reciprocal inductions, or multispecies epigenetics are in his story. . . . Marx came closest in his sometimes lyrical early work. . . . He is both at his most "humanist" and at the edge of something else in these works, in which mindful bodies in inter- and intra-action are everywhere.
> —Donna Haraway, *When Species Meet*

UNTIL RECENTLY, women, and very occasionally men, would make a daily trek to a locked standpipe in the collection of shacks known as Palestine, an area situated within Amaoti, itself a part of Inanda, one of Durban's largest areas of informal housing. The standpipe was operated by a bailiff, who charged local residents some of the highest water rates anywhere in the municipality. This ad hoc arrangement had developed over the previous decade when local councillors struck deals with subcontractors: private fiefdoms of water bailiffs had grown up, tacitly supported by the municipality that provided the potable water, tanked into the standpipe each day. Amaoti is something of a forgotten part of the municipality. A good paved road reaches an end in the informal settlement, stopping abruptly, as if some strange metaphor for the municipality's relationship with the settlement. The area is divided into several different subcommunities, each bearing the name of a real or assumed ally during the struggle years: Moscow, Angola, Cuba, and Palestine. After violent turf-wars broke out across Inanda between the Inkatha Freedom Party and the Comrades movement, the area soon came under the control of the Inanda Marshals—a quasi-militia that sought to keep peace and serve as an interim provider of political representation and services. It is generally accepted that the Marshals acted with remarkable benevolence in the area, dividing up land according to need rather than personal or political gain.[1] After the redrawing of municipal boundaries, Amaoti came under the direct authority of a vastly expanded Durban metropolitan government, something that was later to become the eThekwini Municipality. Changes in the material living conditions for most of the residents were, however, much slower in arriving; hence the situation on February 11, 2003, when, after two weeks in which a water tanker had not appeared at the standpipe, the taps had run dry and local residents had had enough.

Writing on Athens, Maria Kaika describes how when "taps inside people's homes refused to provide their services as expected, they became a form of domestic uncanny: familiar objects which behaved in unfamiliar and disrupting ways. From being invisible and unproblematized, the connections between the house, the city and nature's water became . . . a source of public anxiety and a threat to domestic bliss."[2] Although never a world of domestic bliss, the interruption of water supplies in Amaoti threw the community's daily routines into jeopardy, straining already overburdened bodies and forcing women to travel further afield to find a reliable source of drinking water. The sudden cessation of what was already an unreliable, expensive, and poor-quality supply caused enormous instability within the settlement, exposing the fragility of its relationship with the municipality and the dependence of the politicized environment on the laboring bodies of women. Taps refused to do what they were supposed to do and the injustices created out of this ad hoc water supply were suddenly exposed. On the day of our visit to the area, things came to a head:[3] women were moving from door to door, trying to galvanize the community into a collective response to the crisis. After a few hours, a band of protesters gathered to descend the hill and confront the local councillor. Several hours of intense negotiations, writing of demands, and panicked calls to the municipality took place in her now-locked compound: eventually, in an emotional scene, the protesters were handed the key to the standpipe that had restricted their daily access to water over previous years. An agreement had been made that the tanker would come to the area without charge and that local residents would have control over how water was distributed from the tank. Soon, they were told, a networked supply of water would be brought to the area and the community would receive the same guarantees of a daily supply of water that others in the municipality already had.

"Nature Is Man's Inorganic Body"

For Marx, nature is our "inorganic body." We live from nature and must remain in continuous interchange if we are not to die: "To say that man's physical and mental life is linked to nature simply means that nature is linked to itself, for man is a part of nature."[4] In Inanda, as with any other built environment around the world, various technical and infrastructural accoutrements (taps, tankers, padlocks, buckets) mediate this interchange: their functioning, in turn, depends on a set of institutional, political, and economic relationships. In early February 2003, with the severing of one of these relationships, a limb was temporarily shorn from the community's inorganic body. In response to this local crisis, people were required to suture a new limb that reconnected them to the flows of water composing the municipality. In this chapter, I seek to question how our sensuous engagements with environments shape the possibility for a politicized response to such fractured metabolic relationships. There were many different ways in which the women protesters could have responded—stealing from a neighbor, buying bottled water, giving up hope, trekking even further afield. The response chosen was to reconfigure the relationships constituting the fragile waterscape of Amaoti. Although I refer back to this example in subsequent chapters, here I am concerned with the insights to be gained from Marx's writings on nature, as well as the centrality of nature to his specific vision of the conditions of possibility for a communist future.

As Donna Haraway notes, Marx's writings on nature are at their most lyrically brilliant in *The Economic and Philosophical Manuscripts*.[5] In turn, there can be little doubt that Marx's experiences in Paris in 1843–44—the period in which he was writing the manuscripts—was fundamental to his overall political and economic outlook. For the first time, he was thrown into a community of communist artisans in whose associational practices he saw the seeds for a future society. Capturing these experiences, he writes of how "when communist *workmen* gather together, their immediate aim is instruction, propaganda, etc. But at the same time they acquire a new need—the need for society—and what appears as a means has become an end."[6] The *Economic and Philosophical Manuscripts* distill much of his sense of optimism from these experiences. In particular, they lay out Marx's understanding of alienation under capitalism and his hope for the transcendence of such a fractured life through the practical acts of forging dense associational relationships. Nature is clearly fundamental to this work and to his overall understanding of alienation and a communist future,[7] but it is not always clear how this articulates with the forms of communal practice he so admired at the time. In this chapter, through focusing on our sensuous engagements with produced environments, I will consider how alienation from our inorganic body—nature—is fundamental to the making and remaking of unjust environments within contemporary cities. Going beyond this, I will consider how this deepens the associational possibilities hinted at in the Paris Manuscripts. If, as Haraway suggests, "becoming with" a range of nonhuman others is central to the practice of another globalization, this may also be at the heart of an understanding of the communist hypothesis.[8]

The centrality of nature to Marx's vision of communism is often forgotten.[9] Quite rightly, he is remembered for his optimistic, or, as Benton would term it, utopian challenge to the epistemic conservatism of his predecessors.[10] Reading Marx, this sense of realizable possibilities is captured in the more uplifting moments of his prose: relationships are forged historically; our world remains in flux and open to change; we have the capability to become more complete human beings. In Holloway's terms, our scream at the injustices of the world is never one-sided: it is always one of *both* horror *and* hope.[11] This hope lies in the sense that we need not accept things as they are: hierarchies, gender distinctions, class divisions, poverty, and injustice are not givens; rather, they are products of the splintering and fragmenting forms of social organization we have made for ourselves. If we change the way we organize ourselves and remake our fractured selves, we also transform these crude distinctions. Marx's attacks on Malthus's crude determinism are based on such a liberatory humanism.[12] Nevertheless, if everyday men and women can indeed remake the world in such radically optimistic ways, this seems to leave nature in a somewhat ambiguous position: it can either appear a passive backdrop on which the dramas of human liberation are played out, or, yet more troubling, it can appear as a resource to be dominated and mastered in the quest for technical perfection of the future utopia.

In actual fact, neither of these views comes close to Marx's nuanced materialist outlook, even if the most widely held assumption of historical materialism is that it steers such a productivist and mechanistic line. Instead, nature is more rightly viewed as one active moment in a process of mutual coevolution.[13] The differentiated unity Marx describes both shapes and is shaped by the actions of humans. It both limits and creates conditions of possibility. The realization of a better world, in turn, might transform both "nature" and the senses through which we experience that "nature": "seeing, hearing, smelling, tasting, feeling, thinking, contemplating, sensing, wanting, acting, loving" come to affirm reality in radically different ways.[14] This is the positive humanism Marx develops in his *Economic and Philosophical Manuscripts*, and it is one that continues through subsequent works, albeit implicitly rather than explicitly. Far from a passive backdrop, nature is a key moment in the project of making a better world that contains new associational possibilities with a range of human and nonhuman others.

Following from chapter 1, this insight may seem slightly easier to grasp: for Smith, the production of nature is central to both the production of space and, as I have argued, it can also generate conditions of possibility for a radical democratic politics. Again, however, such insights have been sidelined in favor of a view of Marx as a Promethean who rode roughshod over environmental concerns in a struggle for human liberation.[15] Even within the work of such nuanced theorists as Lefebvre, one reads of how freedom is realized through the progressive domination of nature[16]—an idea also clearly fundamental, although in different ways, to the Frankfurt School's negative sense of the concept of nature in Marx. In this chapter, I seek to develop a very different understanding, thereby contributing to a growing body of literature in which the positive ecological insights of Marx's writings have been brought to the fore. Indeed, a flurry of monographs from the 1990s onward and a series of debates within *Historical Materialism, Capitalism, Nature, Socialism, Monthly Review*, and *New Left Review* have made this task considerably easier. Perhaps the best known and the most enthusiastically received contributions to these debates have come from John Bellamy Foster and Paul Burkett.[17] Rich in detail and seeking to build directly on the *approach* of Marx, such works have deepened our understanding of the role of metabolism and the centrality of ecology to Marx's revolutionary theory. Foster, for example, develops a meticulous reconstruction of the diverse intellectual and material context that shaped Marx's specific approach to materialism. Here he pays particular attention to the influence of Epicurus and, later, emerging debates within soil science, focusing on the specific contributions of Justus von Liebig. From this, Foster develops Marx's understanding of metabolic rift. While drawing on these deeply influential works, I will move in somewhat different directions. In short, I am less convinced of the centrality of ecological crisis to Marx's overall understanding of the contradictions of capitalist societies. Nor am I convinced of the overall importance of the theory of "metabolic rift" to a radical politics of contemporary urban environments. Foster's rediscovery of the roots to

Marx's materialism leads to a neglect of key critiques of mechanistic materialism in the writings of Lukács and Gramsci. Even more curious, Foster then condemns these authors as "idealist" and lacking the coevolutionary perspective necessary for a progressive ecological politics.[18] These criticisms are unfounded, as I will show in later chapters. Indeed, they represent a trend within Foster's work to understand materialism through a narrowly defined lens. For both Gramsci and Lukács, subject–object interrelationships and a dialectical, coevolutionary approach to nature are seen to be fundamental to Marx's explicit and implicit philosophy of praxis.

In this regard, as Haraway suggests, paradoxically Marx's more explicitly nature-sensitive writings emerge in his early works (again, even though giving some weight to the Paris Manuscripts, Foster dwells at most length on several passages from volume 3 of *Capital:* it is on these that he builds his theory of metabolic rift): here Marx presents a vision of nature as something performed through the inter- and intra-action of sensuous bodies. As Foster also shows, this is indeed fundamental to Marx's materialism and to his and Engels's ensuing transcendence of the materialism of Feuerbach and the idealism of Hegel. In this chapter, I wish to take the incipient historical materialism of the *Economic and Philosophical Manuscripts* and the *Theses on Feuerbach* as the basis for thinking through an interventionist and sensuous engagement with urban environments.

In focusing on Marx's early writings, I am not making a distinction between a so-called early and a so-called late Marx. In the 1950s and 1960s, such delineations became the source of heated controversy as theorists such as Louis Althusser and Lucio Colletti claimed a sharp difference between a prescientific humanism in the early writings and a scientific, antihumanist critique of capitalism that only begins to emerge with the materialist view of history found in *The German Ideology*. While emphases and language do indeed shift between the *Economic and Philosophical Manuscripts* and *Capital,* this is not the same as a rupture. Interestingly, while denying a rupture between an early and a late Marx, much of the writing on ecological marxism from Schmidt onward has sought justification for a particular perspective within the "mature approach" developed in *Capital;* it thereby devalues the earlier writings. In this chapter, I argue that if we are serious about the claim that Marx's writings are an integrated whole (in which emphases and language shift even if the overall approach and ideas remain consistent), then the *Economic and Philosophical Manuscripts* are very much worth revisiting. With Lefebvre, I suggest we might come to understand *Capital*—and Marx's overall approach to questions of nature—better through these early writings.

I begin by laying out Marx's concept of nature and will build my argument from there. My thesis is heavily influenced by the reading of Smith in chapter 1, but I bring it into closer conversation with the writings of prominent ecological marxist theorists. Smith does something quite different from other writers, but he includes little of the sensuous embodied ways in which nature is performed—to use a term now prevalent in the literature. Thus, I move on to discuss the importance of the senses to Marx's practical materialism. Andrew Feenberg has already hinted at such a connection in his suggestive claim that the senses are fundamental to Marx's claim that nature is produced through human activity.[19] Although differing in important ways from Feenberg's broader argument, I argue that this opens up new possibilities and ways of bringing Smith's work into conversation with more recent writings in which it is argued that nature is performed. This in turn also underlines Feuerbach's fundamental influence on Marx, albeit one he would subject to critique in subsequent works. Following this, I turn to two contrasting studies within my own work. The first is that with which I opened: here sensate engagements with the degraded environments of a postapartheid informal settlement are crucial to the conditions of possibility for radical political change. The second focuses on creative practice in an effort to prize open new ways of thinking about the experience of the city. Both examples suggest new possibilities for transforming the environment of the city.

The Concept of Nature in Marx

If nature is our "inorganic body," a differentiated unity or an integrated socionatural whole, the constitution of this is, in part, dependent on the creative acts of everyday men and women. Marx was clearly indebted to both Hegel and Feuerbach (and, as Foster demonstrates, Epicurus) in developing such a conception. It is in his early writings, in particular the *Economic and Philosophical Manuscripts* and the *Theses on Feuerbach*, that these debts are most explicit. For Hegel, alienation is a result of the negation of Spirit. Although nature consists of the practical activity embodied in Spirit and is, necessarily, an integral part of the external world, alienation prevents consciousness of this totality. Through acquiring consciousness of the unity of Spirit and the external world, alienation is overcome. In short, the becoming of Spirit reunites Man and Nature in a conscious unity. The opposition of subject and object, of humans and nature, is negated through Hegel's dynamic view of history.

Although logically sound, this movement demonstrates the crucial difficulties that a conception of nature poses for a rigorous idealist philosophy.

Nature in Hegel appears to be dissolved into History, and yet nature is surely always that which is unmediated, that which remains constant outside the transformations of society and history: this was the position Kant reached, in which a distinction between the phenomena and the noumena, or "thing in itself," was preserved for the sake of logical consistency. As Lukács argues, Kant was at least honest in presenting the problem in its clearest way, even if he was not able to overcome the resulting antinomy.[20] If Hegel overcame such a problem, it was by dissolving nature into the movement of the dialectic. This represented a form of idealism that Marx and Engels were to consistently reject. Directing their frustration against such fanciful views as they came to be represented among the Young Hegelians, Marx and Engels claim:

> Once upon a time a valiant fellow had the idea that men were drowned in water only because they were possessed with the idea of gravity. If they were to knock this notion out of their heads, say by stating it to be a superstition, a religious concept, they would be sublimely proof against any danger from water. His whole life long he fought against the illusion of gravity, of whose harmful results all statistic brought him new and manifold evidence. This valiant fellow was the type of the new revolutionary philosophers in Germany.[21]

Historical materialism confronts such idealism through foregrounding practical, creative activity in forging the metabolic interchange between humans and their environments. This was the point at which I concluded the last chapter, arguing that while not determinant, practical activity retains a specific importance within historical materialism. Hegel's laboring of consciousness is superseded by the concrete acts of laboring individuals within historically specific social structures.

Surprisingly, it is in *The Economic and Philosophical Manuscripts* that we find some of the most detailed discussions of this transformation of the idealist dialectic. Clearly influenced by Feuerbach as "the only person who has a *serious* and a *critical* attitude to the Hegelian dialectic and who has made serious discoveries in this field,"[22] Marx goes on to criticize others among the Young Hegelians who merely used philosophy to negate theology while returning once more to theology. All except Feuerbach neglected the importance of the real relation of "man to man." This is the basic principle of Feuerbach's philosophy, hence his claim in *Principles of the Philosophy of the Future* that "the true dialectic is not a *monologue of the solitary thinker with himself.* It is a *dialogue between 'I' and 'You.'*"[23] Building on this principle, Marx was to argue that the intervention of Spirit in Hegel's concept of nature is itself one of the most profound forms of alienation and one that Hegel is never able to overcome:

subject–object unity is achieved for Hegel only in the service of Spirit. For Marx, Hegel recognizes the world only as an objectified form of thought whose purpose is to promote the self-realization of the Idea.[24] Rejecting such objectifying thought as yet another form of alienation, Marx frees the dialectic of this external intervention. Hegel's radical vision only achieves the self-realization of the Idea in consciousness and not in the still-inverted reality of capitalist society. Contrasting his own vision, Marx argues that it is through the concrete, sensuous acts of everyday people that alienation will truly be challenged. His vision of communism is thereby based on achieving such a nonalienated unity of the social and natural in sensuous activity. This is utterly exemplary of his philosophy of praxis: the philosophical conundrums posed by his German Idealist predecessors find their resolution in the concrete acts of everyday men and women.[25] At the same time, the philosophical categories through which the Idealists had viewed the world are raised to the position of material concepts. Subject and Object are found to be people and things rather than mental fancies.

Alienating Metabolisms

Marx reaches such a conclusion through a dialectical understanding of another key Hegelian concept, that of alienation. Prior to the resolution of subject and object through Spirit coming to know itself as absolute reality, there is a separation embodied in the alienated Spirit. Again, this emerges through a lack of consciousness of the fact that Spirit and the external world are one. Within Marx, alienation takes a form specific to the type of organization within capitalist society. Crucial to this is the separation of the worker from his or her product and from the conditions of production in which he or she works. Here, as Foster and Mészáros show, the alienation discussed in the *Economic and Philosophical Manuscripts* is inseparable from alienation from nature.[26] It is important to consider this, for in it lies the conditions of possibility for a radically reorganized set of political ecological relationships.

As discussed in chapter 1, Marx saw the metabolic interchange between society and nature as a mutually determining one. Levins and Lewontin capture this in more strictly Darwinian (and yet still deeply marxist) terms:

> The incorporation of the organism as an active subject in its own ontogeny and in the construction of its own environment leads to a complex dialectical relationship of the elements in the triad of gene, environment and organism. . . . The organism is, in part, made by the interaction of the genes and the environment, but the organism makes its environment and so again participates in its own

construction. Finally, the organism, as it develops, constructs an environment that is a condition of its survival and reproduction, setting the conditions of natural selection.[27]

To take a different example, and one I am perhaps in danger of romanticizing, I think of my brother, a boatbuilder, imparting aspects of his own creativity into the boat on which he is working. At the same time, he learns from the experience, figuring out how to fit decking in the most efficient and elegant fashion, while also gaining calloused hands, a bad back, and perhaps on occasion, a sense of joy from the process. In turn, the boat he produces enables a new set of relationships to be forged by those sailing it. The relationship is a reciprocal one between the laboring subject and the object being created. Nevertheless, capitalism transforms the directness of this reciprocal process, driving subject and object apart. The boat embodies both a use value—its ability to carry people, bring joy, and connect places—as well as an exchange value. More and more, exchange value comes to dominate within the production process (even if use value remains fundamental). Thus, as workers produce commodities within capitalist society, they are not producing for themselves or for those with whom they have any direct connections. Instead, goods are produced for exchange. Whereas concrete labor continues as the process through which raw materials are transformed into commodities, in the historically and geographically specific conditions of capitalist societies, abstract labor comes to be the measure of this process. This is implied in Marx's distinction between concrete and abstract labor and in his modification of Ricardo's labor theory of value by introducing an understanding of "socially necessary labour time."[28] Deskilling and the dominance of this abstraction in both the workplace and the home is a common feature of this shift. As a boatbuilder, my brother negotiates between these positions, seeking to retain his role as a craftsman while surviving in a viciously competitive market.

The dominance of abstract labor implies a fundamental change in the sensuous relationship to nature. Smith captures this fundamental shift in the manner in which first nature and second nature are distinguished not in terms of one being produced and one not (concrete labor remains as fundamental as ever), but through second nature being dominated by the exchange relation: "The distinction now is between a first nature that is concrete and material and a second nature which is abstract, and derivative of the abstraction from use value that is inherent in exchange value."[29] Although still embodying aspects of our personality, things produced under capitalist conditions become entities that are radically divorced from us:

This fact simply means that the object that labour produces, its product, stands opposed to it as *something alien*, as a *power independent* of the producer. The product of labour is labour embodied and made material in an object, it is the *objectification* of labour. The realization of labour is its objectification. In the sphere of political economy this realization of labour appears as a *loss of reality* for the worker, objectification as a *loss of reality* for the worker, objectification as *loss of and bondage to the object*, and appropriation as *estrangement*, as *alienation* [*Entäusserung*].[30]

If we take Smith's claim that nature is produced and that, under capitalist conditions of production, this becomes a specific type of product that is abstracted from use value,[31] then the claim that alienation from nature is fundamental to Marx's vision of communism makes more sense. Without this understanding, it is very difficult to reconcile Foster's enthusiastic agreement with Marx's claim that nature and society are fundamentally interwoven and the claim that humans are alienated from this.[32]

Indeed, Foster's development of the theory of metabolic rift, in which Marx argues that the division between town and country, as well as the increasing commercialization of fertilizer, ends the crucial organic link between humans and the earth as human waste no longer feeds the soil is one of the few examples where this seems possible. However, the implications of Foster's thesis for contemporary thought are vague and the conclusions atavistic.[33] Is Foster really advocating a society in which night soil is routinely used from major cities to fertilize the agricultural needs of the urban populace, thereby reconstituting the alienated relation with the land severed by urbanization and capitalist farming practices? Harvey comes closer to a workable understanding of the ecology of modern cities—and one that reflects a far more interesting application of notions of alienation and metabolism when he writes of how

> we cannot somehow abandon in a relatively costless way the immense existing ecosystemic structures of, say, contemporary capitalism in order to "get back close to nature." Such constructed ecosystems are a reworked form of "second nature" that cannot be allowed to deteriorate or collapse without courting ecological disaster not only for the social order that produced it, but for all species and forms that have become dependent on it. . . . The created environments of an urbanizing world, their qualities and particular difficulties . . . have to move to the centre of our attention relative to much of the contemporary preoccupation with wilderness.[34]

It is in this regard that we might begin to rethink the example of Inanda given earlier. It is not the case that the diverse communities of Amaoti were totally separated from the flows of water that shape the city: life would be inconceivable if this was the case. Everyday life in Amaoti depended on an ongoing interchange with the produced environment of the settlement and the municipality beyond. Purified, treated water was brought by tanker to some parts of the settlement, while others were able to access some form of piped supply. In what sense might we consider this relationship to be an alienating one, then, when there is not the same physical separation from produced nature as that which is implied in Marx's account of production?

Amaoti must be understood as the "created ecosystem" that Harvey discussed. Alienation operates within such an ecosystem, even while humans remain a fundamental part of it. Thus, challenging alienation does not mean establishing a simple direct connection to untreated water as might seem to be the logical correlate to Foster's theory of metabolic rift. In spite of the hopeful promise of intermediate technologies such as rainwater harvesting, to argue for a simple "off-grid" relationship to water within the community would be to condemn Amaoti's residents to a life of disease and misery. In most cities of the world, challenging alienation as it is constituted within water provision cannot simply mean moving to "off-grid" supplies and regaining a direct connection with local watersheds: this, as Harvey implies in the quotation above, would be to court ecological disaster as well as reimpose profound social injustices. Instead, the relationships through which this purified water is distributed might be questioned. This implies a somewhat different case of "alienation from nature" from the way in which it is most often understood. Too often, this would appear to suggest the loss of connection with a preexisting natural world and the need for advanced industrial societies to rediscover such enchantment with nature. In contrast, in the moment of crisis that manifested itself in Amaoti on February 11, 2003, as tankers stopped coming to the community in order to charge exorbitant rates for a substandard service, the relationships that connected the community to water in an abstract manner (based primarily on exchange value) came to the fore. These relationships need to be challenged in various ways: there can be no reconnection with some mythical preexisting, pristine nature. In challenging these relationships, the possibility exists for radically reconfiguring and reconnecting with what Marx terms "species-being."

Species-Being

By species-being, Marx seems to imply that humans have the capability of *becoming* through freely producing together. It distinguishes human from nonhuman, and this distinction, in turn, is related to the production of nature:

The practical creation of an *objective world*, the *fashioning* of inorganic nature, is proof that man is a conscious species-being, i.e. a being that treats the species as its own essential being or itself as a species-being. It is true that animals also produce. They build nests and dwellings, like the bee, the beaver, the ant etc. But they produce only their own immediate needs or those of their young; they produce one-sidedly, while man produces universally; they produce only when immediate physical need compels them to do so, while man produces even when he is free from physical need and truly produces only in freedom from such need; they produce only themselves, while man reproduces the whole of nature; their products belong immediately to their physical bodies, while man freely confronts his own product.[35]

In this passage, humans are to be distinguished from animals by the fact that they produce freely and universally rather than for the specific needs of themselves or their offspring. Species-being is thereby reflected in the production of nature. Nature embodies and expresses the species life of different men and women brought together in a common unity. Within capitalist societies, however, this species life is alienated from people, it "degrades man's free activity to a means, it turns the species-life of man into a means for his physical existence." In short, it transforms humans to animals and renders nature exterior to humans. Through the concept of species-being, Marx is asserting a fundamental unity of nature and society, albeit one that is presented through a series of antitheses—"them" and "us." However, crucially, within capitalist societies, this unity has been wrought asunder by the abstraction of labor.

Needless to say, few concepts have been so universally derided within Marx's oeuvre as that of species-being. Commonly, species-being is seen as an overt example of essentialism, in which Marx is asserting a universal condition for humans that has been perverted by the workings of capitalist society. This, many concur, is humanism at its worst. Presented as that which makes humans distinct from animals, it seems to assert the very binaries that are so problematic for an environmental politics. Nevertheless, although abstract, species-being need not define an essential condition; read through other writings, the implication of the concept is that we might explore new forms of coevolution or new forms of "becoming with as a process of becoming worldly," as Haraway might have it. It is for this reason that authors such as Harvey refer to species-potential rather than species-being. This is the same point as that made with regard to alienation: there is not a route "back" to some pre-existing relationship with nature to be found in Marx; instead, we need to rethink both contemporary and historical relationships with nature as a way of reworking these relationships in the future. (Perhaps here too the nonlinear materialism of Epicurus referred to in Foster's work is also exerting its influence on the Paris Manuscripts).[36] In short, the criticisms of species-being that abound are indeed perfectly valid. But the concept referred to is not what Marx appears to develop in *The Economic and Philosophical Manuscripts*. Here there is a far more nuanced sense of the potential for a creative and symbiotic relationship between human and nonhuman to be achieved through sensuous engagement in the present and for the future. This is not to justify the loaded valuations of human and animal that do emerge within the *Manuscripts*; rather, it is to draw attention to the coevolutionary perspective that remains at its heart.

Finally, bringing together the previous manifestations of alienation, Marx argues that alienation separates individuals from one another. He then goes on to consider the reasons for this self-alienation, posing the question: "If my own activity does not belong to me and is an alien, forced activity, to whom does it belong then?" Given the incipient philosophy of praxis that Marx develops here, the answer can only be that this activity belongs to other humans. Alienation is neither produced by gods nor nature: it is a product of the ways in which humans have organized their activity, reliant as this is on a separation of the producer from his or her product. Taking this position forward in the *Grundrisse*, Marx reframes the question: "It is not the *unity* of living and active humanity with the natural, inorganic conditions of their metabolic exchange with nature, and hence their appropriation of nature, which require explanation or is the result of a historic process, but rather the *separation* between these organic conditions of human existence and this active existence, a separation which is completely posited only in the relation of wage labour and capital."[37]

Smith clearly recognizes the importance of this enforced separation of the producer and product. Thus, he writes of how, in contradistinction to the domination of nature thesis, Marx "begins with nature as a unity and derives as a simultaneously historical and logical result whatever separation between them [nature and society] exists."[38] However, in seeking to assert that nature is indeed a differentiated unity, Smith does not dwell at all on the historical conditions that have achieved a separation at key moments. Through considering in more depth the role of alienation in Marx's ecological politics, I think we gain a clearer sense of this. Alienation depends on the production of nature: it is concerned precisely with the (historically understood) metabolic process through which humans and socio-natures co-constitute one another. Within capitalist societies, the result of this is an inversion. A nature is produced that is radically divorced from the society of which it remains a part. This product of sensuous activity appears as an alien force. Such a position seems utterly contradictory: nature, on the one hand, is a socio-natural entity and part of human society, and yet, on the other hand, it is something alien to those who are an integral part of it. This, I would suggest, is at the heart of some of the confusion of Marx's view of nature. It can only be grasped by thinking through the consequences of alienation and how this emerges from and is integrally related to socio-natures.

For McLellan, whereas the *Manuscripts* comprise Marx's negative assessment of the fate of the world in capitalist society, his writings in a separate notebook and under the heading "On James Mill" represent his positive view of the reconciliation of people with the object of their labor, with the work process itself, with their species-being, and with one another.[39] There are clearly aspects of this within the *Economic and Philosophical Manuscripts*, too. As asserted in the epigraph with which I began this chapter, a true reconciliation with nature is integral to Marx's view of communism. He distinguishes this from several other crude formulations of the time. For Marx, human potential can only be realized and a genuine reconciliation of nature and society achieved through the abolition of wage labor. This will dissolve the separation between worker and product while confronting the true sources of the dualisms on which capitalist society rests. It does not take much to detect within Marx's vision of communism a sense of artistic practice as central.[40] He seems to have a particular model at the heart of his view: creativity is stunted under capitalism and creative fulfillment will only be achieved in a future communist world. In turn, this conception of communism depends on reuniting sense perception with creative activity, such that the senses find themselves confirmed rather than alienated in creative practice.

The Senses as Directly Theoreticians in Practice: Sensuousness and the Creative Act

Throughout the *Economic and Philosophical Manuscripts*, Marx returns to the possibility of the senses becoming "directly theoreticians in practice." If practical activity is at the heart of the process through which nature is humanized and humans are naturalized, sensuousness is fundamental to this active, practical materialism. In turn, the senses themselves are understood to be shaped relationally and historically: they are one further moment in the process of mutual coevolution already described. This is a radical claim, and it is worth quoting a passage at length. Here Marx begins to mobilize an artistic model, arguing for the centrality of the senses in both a critique of the existing world and in the forging of a future communist world:

> Only music can awaken the musical sense in man and the most beautiful music has *no* sense for the unmusical ear, because my object can only be the confirmation of one of my essential powers, i.e. can only be for me in so far as my essential power exists for me as a subjective attribute (this is because the sense of an object for me extends only as far as *my* sense extends, only has sense for a sense that corresponds to that object). In the same way, and for the same reasons, the *senses* of social man are *different* from those of non-social man. Only through the objectively unfolded wealth of human nature can the wealth of subjective *human* sensitivity—a musical ear, an eye for the beauty of form, in short, *senses* capable of human gratification—be either cultivated or created. For not only the five senses, but also the so-called spiritual senses, the practical senses (will, love, etc.) in a word, the *human* sense, the humanity of the senses—all these come into being only through the existence of *their* objects, through *humanized* nature. The *cultivation* of the five senses is the work of all previous history.[41]

Building on this relational understanding of the senses, Feenberg claims that truth is not to be found in a world beyond sensation but in truer and deeper sensation. Transformative change will alter not just the objects of the senses but the senses themselves, such that "revolution unites subject and object in liberated sensation and thereby reveals the truth of nature."[42] Here the symphony, the sonnet, and the landscape painting are seen as one aspect of the "humanization" of nature in which humans themselves are also shaped by the sensuous objects (or produced nature) with which they are engaging. Although in no way succumbing to a simplistic environmental determinism, Marx accords nature, and the environment in which humans live, with an agency not often associated with historical materialism.[43] Marx clearly has a model of artistic practice in mind. Just as a musical ear is trained by the objects with which it is engaging, so a broad range of sensual experience will be transformed through the achievement of a new metabolic relationship with the material world following a communist revolution. Later, this was to be picked up in different ways by Lukács, who relates this conception of nature to a mode of artistic practice, claiming that nature is: (a) that which we are not; (b) that within us; (c) that which we might become, seen in the principle of creation in art.[44] He continues:

> Art shows us ... the two faces of Janus, and with the discovery of art it becomes possible to provide yet another domain for the fragmented subject or to leave behind the safe territory of the concrete evocation of totality and (using art at

most by way of illustration) tackle the problem of "creation" from the side of the subject. The problem is then no longer as it was for Spinoza—to create an objective system of reality on the model of geometry. It is rather *this* creation which is at once philosophy's premise and its task. This creation is undoubtedly given ("There are synthetic judgements a priori—how are they possible?" Kant had once asked). But the task is to deduce the unity—which is not given—of this disintegrating creation and to prove that it is the product of a creating subject. In the final analysis then: to create the subject of the creator.[45]

This analysis finds its fulfillment in the landscape concept, even if, as Lukács goes on to argue, it in no way solves the antinomies of bourgeois philosophy: the art object in this case remains an object. Lefebvre, however, takes this understanding of creative practice forward by extending it to the realm of everyday life. "Let everyday life become a work of art" shouts from the pages of his critique of everyday life:[46] here the senses are transformed from theoreticians that confirm the separation of human praxis from material conditions of existence to an affirmation of a better life in which the act of creation is fundamental to the most banal exchanges. I will return to these points in more depth in chapters 3 and 5.[47]

Perhaps in part because of this crucial privileging of artistic engagement with the world, musical allegories have been used throughout the literature on Marx's concept of nature. Feenberg writes of how "In alienated society man experiences nature as a dog or cat might experience a symphony" and "as the 'musical ear recognizes itself 'affirmed' in the music it hears, so will liberated humanity recognize itself affirmed in nature."[48] For Grundmann, a musical allegory provides a possibility for transforming crude interpretations of the mastery of nature: "Imagine a musician who plays her instrument with virtuosity. We call her play 'masterly.'. . . It is in this sense that we have to understand the domination of nature. It does not mean that one behaves in a reckless fashion towards it, any more than we suggest that a masterly player dominates her instrument (say a violin) when she hits it with a hammer."[49] Harvey appears to concur, albeit while criticizing the gendered overtones.[50] Nevertheless, Benton, building on his overarching call for a reconstruction of marxist theory on the basis of a recognition of "natural limits," argues that the allegory is misplaced: he writes that "I would myself be inclined . . . to speak of mastery of the skills, or techniques, of violin-playing, rather than of 'mastering the violin'" and "The violin is not appropriated or transformed as it is played."[51] Feenberg concurs, writing that "the universe is not, in principle, mere raw material for labor: the very idea is either absurd or abhorrent."[52]

I am less interested in the debate over whether or not "mastery" implies domination (or indeed whether domination itself has been misconstrued).[53] More interesting is the manner in which the production of nature outlined in the previous chapter can be understood as a sensuous process that mirrors aspects of the act of artistic creation. Indeed, Feenberg's rejection of what might be termed a "strong" version of the "production of nature" thesis (perhaps best represented in Smith's work, although not referred to by Feenberg) is made possible through revisiting the radical development of a theory of sensation—itself related to modes of artistic practice—within Marx's writings. Here is Feenberg:

> The senses, unlike labor, have traditionally been conceived by philosophy as a potentially universal mode of reception, relating to all possible (real) objects. The senses can therefore take over where actual labour leaves off, supporting the assertion of a universal identity of subject and object in nature.[54]

And again:

> The emancipated senses are active transformers of their objects and not mere passive receptors; they can be understood on the model of the labour process as engaging in a theoretical-practical activity, objectifying human nature and releasing the implicit potentialities of the material on which they work.[55]

In making this claim, Feenberg rejects the fall-back position of both Habermas and Schmidt that Marx requires a Kantian resolution to the seemingly impossible claim that all nature is produced through human practice (see also, for a radically different argument, Smith's rejection of this in chapter 1).[56] It is, he argues, in the new theory of sensation that the radical claim that nature is produced might be seen to be workable. On several counts, I would argue that Feenberg is wrong. His claim that "the universe is not, in principle, mere raw material for labor: the very idea is either absurd or abhorrent,"[57] is to distort the claims being made by Marx. His attempt to rescue Marx from either absurdity or mixed metaphor, culminating in a "unique form of phenomenalism," appears to take us far from the intent of Marx's writings.[58] Nevertheless, even if ultimately missing the point, Feenberg's emphasis on the centrality of the senses is utterly lacking in Smith's otherwise brilliant analysis. In neglecting this, he occludes the embodied ways in which these coevolutionary processes emerge as sensuous praxis. Clearly, the roots to such a conception must lie in Marx's engagement with both Epicurus and Feuerbach.[59]

Marx's debt to Feuerbach is frequently understood to lie in the latter's critique of religion. Turning on its head the claim that God created the universe,

Feuerbach was to argue that God is in fact a projection of the thought of Man: "God is nothing else than Man: he is, so to speak, the outward projection of Man's inward nature." Feuerbach's critique of religion is materialist in the sense that it challenged the idealist philosophies that preceded it by placing creative energies in the hands of Man. Nevertheless, such an atheistic critique would have been commonplace to Marx. Indeed, the Young Hegelians composed a group built on such critiques of religion: Marx's true debt to Feuerbach—and his shortlived but intense enthusiasm for his work—must surely lie elsewhere. In particular, Marx appears drawn to Feuerbach's repositioning of sensuousness as fundamental to a philosophy of the future. Such a position becomes clear in the *Theses on Feuerbach*, a work in which Marx's debt to Feuerbach and also his transcendence of both Feuerbachian materialism and Hegelian idealism is at its clearest. Never intended for publication the *Theses on Feuerbach* was reportedly posted above Marx's writing desk as a note to himself. They capture his emerging practical materialism superbly and lay out much of the groundwork that was later to develop into his collaboration with Engels in *The German Ideology*.

In the first of these *Theses*, Marx praises Feuerbach for grasping "sensuous objects, really distinct from thought objects."[60] For Feuerbach, the human essence is confirmed through sensuous objects. The eye finds its essence not in the fact that an object can be heard but in that it can be seen. It finds its essence in the blue sky, the golden leaves, or the vivid oil painting. The new theory of sensation developed in *The Economic and Philosophical Manuscripts* had already emerged from such sensationalism. For Marx, however, he was soon to be frustrated with the lack of movement in Feuerbach's philosophy—the missing sense of active interpenetration of subject and object—in Feuerbach's claims. Nature, for Marx, was something undergoing constant transformation, and so, if the human essence is to find its confirmation in nature, it too must be transforming. Thus, completing the first Thesis on Feuerbach, Marx moves on to criticize Feuerbach, "but he does not conceive human activity itself as *objective* activity. . . . Hence he does not grasp the significance of 'revolutionary,' of 'practical-critical,' activity." Again, this practical-critical activity is a deeply sensuous experience and confirms the potential role of the senses "as directly theoreticians in practice." This also explains the claim by Marx and Engels in *The German Ideology* that "so much is this activity, this unceasing sensuous labour and creation, this production, the basis of the whole sensuous world as it now exists, that were it interrupted only for a year, Feuerbach would not only find an enormous change in the natural world, but would very soon find that the whole world of men and his own perceptive faculty, nay his own existence, were missing."[61] As the *Theses on Feuerbach* progresses, Marx goes on to: call for a decidedly practical materialism that might be distinguished from scholastic practice; criticize the naive assumption that an unthinking populace merely need to be educated to change the world ("it is essential to educate the educator himself"); and, particularly important, criticize Feuerbach for his abstract and reductive notion of Man ("the human essence is no abstraction inherent in each single individual. In its reality it is the ensemble of the social relations"). Later, as we will see, such historical and relational understandings are taken up within Gramsci's *Prison Notebooks*, where the true radicalism of the *Theses* is really exploited.[62] Here a relational understanding of human nature is transformed into a fluid, coevolutionary understanding of what might be termed socio-natural relations. In short, through situating Marx's radical theory of sensation within a broader understanding of his engagement with Feuerbach, we can see how a dynamic, historically contingent view of socio-natural relations begins to emerge. Herein lies an emerging critique of universalist or "naturalizing" assumptions regarding history, geography, and the production of space. The senses are crucial to this, mediating the humanization of nature and the naturalization of humans. Perhaps surprisingly, in spite of its avowed humanism, the position arrived at here is not radically dissimilar from some work in what might be termed the posthumanities.[63]

Sensuous Engagements with Metropolitan Nature

In recent years, sensory experience has become fundamental to a range of interventionist artistic practices. This serves as one laboratory in which a radical theory of the senses has come to be experimented with. Several interventions serve as invitations to rethink the city as a mosaic, one that is confirmed by—and gives confirmation to—our own sensuous experience. This appears to echo Feenberg's interpretation of Marx. If we apply it to a stronger understanding of the production of nature, it also provides ways into thinking about the political possibilities for reconfiguring urban environments. Providing some context to recent critical spatial practices, Jane Rendell argues that they have emerged out of a critique of existing public art projects: here she quotes Malcolm Miles on how public art can serve to provide wallpaper over existing social conflict, as well as promoting key corporate interests and dominant ideologies. She goes on to describe how critical spatial practices occupy an interdisciplinary terrain "seeking to make a new kind of relationship between theory, specifically critical theories that are spatial, and art and architectural practice."[64] She then goes on to specify this more clearly in relation to site-specific artwork, the notion of the expanded field and the production of space. Miles is perhaps more concrete, in writing of urban interventions as artistic projects that seek to work as both provocations and participatory entry points for new forms of collaborative artistic practice.[65] It is helpful to think through some of the sensory experiments in these terms.

Perhaps one of the more interesting examples emerges in the work of Christian Nold, who has focused on developing what he terms *Biomapping*.[66] The idea behind *Biomapping* is quite simple. Participants wear both a lie detector and a Global Positioning System before going for a walk within a specific neighborhood. The lie detector picks up on sensory stimuli that evoke a particular response—perhaps stress or excitement, joy or panic. Through the Global Positioning System, these stimuli are then mapped, usually being represented by a line graph on which the peaks and troughs reflect a stimulus or otherwise. Coloring can add to the overall representation. The combined effect of different participants performing the same task, albeit via different routes, and clearly experiencing environments in different ways allows a composite "emotion map" of a neighborhood to be developed.

Nold's *Biomapping* is, however, far from an unambiguous tool for liberation. Its translation of the richness of human experience to a digitally mappable form can appear crude and potentially open to capture by powerful interests. The aim, nevertheless, is to transform technologies that have been used for domination into democratic resources for experiencing the environment of an area in radically different ways. In this regard, it poses questions about the links, if any, between technology, emotion, democracy, and liberation. Even though Nold would deny the connection, there are links to the work of the Situationists and their strategies of *dérive*, targeted at developing new ways of experiencing familiar neighborhoods as a tactic for undermining the power of the Spectacle. This in turn has many positive links to both Lukácsean understandings of reification and the writings of Marx discussed in this chapter. However, what interests me here is the potential for such work to provoke consciousness of the myriad sensuous and embodied experiences constituting the urban environment. Our senses are in part shaped through living, working, and playing in cities. We drown out certain sounds in order to focus on others. From Simmel onward, there has been recognition of the intense stimulation of the senses in urban contexts. But as Nold's work shows, these sensory experiences are then imparted into the very fabric of the city, even if we cannot always see this process. The urbanization of nature, as Swyngedouw and Kaika refer to it, is a profoundly sensual experience and the resultant environment is a sensory mosaic.[67] There is a mutual exchange between the material environment of the city and the sensory experience of those living within it. In The Newham Sensory Deprivation project, Nold worked with sixth-form students in order to demonstrate how, in "depriving" participants of their senses of sight and sound, radically different emotion maps emerge. Although a cruder example, this further draws out the mutual interchange between subject and object in the making of urban environments.

In a somewhat different vein, and directing their attention toward a more explicit environmental politics, the artistic collective Proboscis has pioneered a series of collaborations with various computer scientists in order to develop what they term "robotic feral authoring."[68] Remotely controlled robots, fitted with a variety of devices to measure air and sound pollution, are sent on missions to explore London streets or parks. The information they acquire is transferred to a map to demonstrate aspects of the environment missed by others. Working with various London schools, Proboscis seeks to develop new methods that permit children to map the ambient environment around them.

Although neither of these projects achieves a radically different production of urban environments, each draws attention to the processes, emotions, gases, and relationships that do produce metropolitan nature. In this sense, they create conditions of possibility through which the urbanization of nature and the making of the person in relation to it might be rethought. This can be considered in a particularly functionalist way, and both Nold and Proboscis have been commissioned by local authorities interested in the potential planning applications of their work. If a particular road junction generates enormous stress, perhaps its layout needs to be reconsidered, for example, and a safer road crossing could be installed. Or, if a corner of a London park shows particularly high carbon monoxide concentrations, this is clearly not the corner in which to install a children's play area. However, I am less interested in these applications and far more interested in the ways in which representations of the urbanization of nature as a sensuous process might lead to the kind of politicized response and the affirmative politics that emerged in the example of Amaoti, the example with which I began this chapter.

On the surface, the anger of residents in Amaoti and the playful interventions of Nold and Proboscis are almost antithetical. One is a politics born out of desperation and anger, the other is a politics emerging from the apparent profligacy of a Northern artistic world in which end points require no predetermination and process-based interventions abound. However, I wish to argue that both situations demonstrate similar conditions of possibility emerging from a politics that builds upon the metabolic exchange with nature in an urban context. In both cases, that metabolic exchange involves a sensuous engagement with a produced environment. The laboring reproductive acts of women in Amaoti produce and reproduce the inorganic body that Marx understands nature to be. As one crucial aspect of this was forcefully separated from the settlement, potable water (one of the more direct forms of a "produced nature") came to be alienated from the lives of those living within it. Again, this was a directly sensed experience—through thirst, stress, anger, and the hot sweaty labor of walking a mile farther to acquire an alternative supply. Countering this alienation required striking at the heart of the relationships through which water was divorced from the community. This took both direct forms—acquiring the key to the locked standpipe in Amaoti—and indirect forms, ensuring that the municipality recognized the citizenship of those within the settlement and thereby ensured the entitlement to universal free basic water afforded to the rest of the city.

If the work of Nold and Proboscis develops a more playful politics, it is nevertheless one utterly premised on the metabolic exchange with nature that shaped the radical intervention of women in Amaoti. Urban metabolisms are more often than not either taken for granted or unnoticed. Bruno Latour and Emilie Hermant's richly textured *Paris: Invisible City* draws attention to the hidden networks out of which Paris is produced as a created ecosystem or socio-natural assemblage.[69] Although not referred to as such, Nold's work uncovers the sensuous exchanges out of which the city is created: the fears, the shocks, the joy, and the anger. Grasping a politics of everyday urban natures involves working with these sensuous exchanges and recognizing that these are the very conditions of possibility for a radical socio-natural politics.

Conclusions

The wealth of important studies of Marx's concept of nature in recent years has made it far harder to make the unsubstantiated claim that Marx's historical materialist approach is somehow anti-ecological, Promethean, or mechanistic. Several authors have sought to demonstrate how a form of ecological crisis was central to both Marx's materialism and to his critique of the contradictions of capitalist society. In this chapter, I have sought to add one further layer to this understanding by focusing more on the forms of relational sensuousness so fundamental to Marx's vision of a nature produced under the alienating conditions of contemporary capitalism and of the possibilities for transforming this in a future society.

If the two studies from which I have drawn appear wildly different, this is in part my intention. For both demonstrate differing but related potentials for a future politics. Such a politics must reject the forms of injustice out of which socio-natures are alienated, at the same time as providing an understanding of the sensuous engagements and creative practices out of which city environments might be remade. In exploring these engagements and practices, it becomes clear how central a socio-natural politics and a form of creative practice are not only to Marx's vision of communism but to a form of politics capable of transforming our contemporary cities in positive, progressive ways.

Chapter 3 Cyborg Consciousness
Questioning the Dialectics of Nature in Lukács

> The question to be asked today... of the early Marxist Lukács is not: "How does this work stand in relation to today's constellation? Is it still alive?" but..."what is dead and what is alive in Hegel's dialectic": how do we today stand in relation to—in the eyes of—Lukács? Are we still able to commit the act proper, described by Lukács? Which social agent is, on account of its radical dislocation, today able to accomplish it?
>
> —Slavoj Žižek, "Postface: Georg Lukács as the Philosopher of Leninism"

> Cyborg consciousness can be understood as the technological embodiment of a particular and specific form of oppositional consciousness.
>
> —Chela Sandoval, "New Sciences: Cyborg Feminism and the Methodology of the Oppressed"

RETURNING TO THE SITUATION in Amaoti referred to in chapter 2, there seems to be something about the sensuous laboring acts of women that generates conditions of possibility for the radical response to the cessation of water supplies that was witnessed. This should not be romanticized. Rather, a profoundly gendered division of labor ensures a distribution of tasks that is generally favorable to men and, in turn, ensures that most men and most women in Amaoti interact with environments in profoundly different ways. Central to the perspective that guides this book is a belief that world-changing perspectives emerge from consciousness of the coevolutionary processes out of which environments are constituted. Given this, the conditions of possibility for a radical consciousness of these (gendered, raced, and classed) technonatural environments are clearly worth exploring.

Making the daily trek to a standpipe to fill a plastic bucket with water mediates a set of relationships with structures of local government; with distant shareholders and investment bankers; with gendered, raced, and classed subjects; and with the ground- or surface-water sources from which the supply is provided. Nevertheless, the organization of technonatural networks, not to mention the apparent artificiality of cities, often leads to dualistic forms of environmental thinking in which nature is divorced from society. Anger turns not to the relations barring access to water but rather to the water meter itself or to technocratic solutions that merely limit supplies to the most needy. In this chapter, I will argue that gaining consciousness of these mediations is fundamental not only to ensuring democratic access to supplies of water but also to the conditions of possibility for a politicized understanding of the environment.

Although far from an unproblematic contribution, this is the central thesis of Georg Lukács's magnum opus, *History and Class Consciousness*. For many, *the* most important text in marxist philosophy, this work prioritizes a Hegelian methodology in Marx's writings, claiming that it is the proletariat who is the subject–object of history and whose destiny it is to change the world. Through a unique ability to acquire consciousness of "the totality," the proletariat comes to represent for the early Lukács what Michael Löwy quite rightly terms the "messiah class."[1] This is a *revolutionary humanism* of the highest pitch: here, the acquisition of consciousness by the working class fundamentally changes existing reality. Through this radical claim, Lukács's early writings have quite rightly been seen to transform the direction of marxism and science and technology studies, as well as being of fundamental importance to many of the most important threads of feminist standpoint theory. In what follows, I wish to consider Lukács's central arguments in the wake of posthumanist concerns with the socio-natural. For many, there is both a deep-seated arrogance and a dangerous speciesism within the humanism developed by someone such as Lukács. The purpose of this chapter is therefore to appropriate the radicalism of this critique of capitalist society, as well as the sense of immanent possibilities within everyday life, while bringing this into conversation with more recent disruptions of the human exceptionalism underlying such claims. I remain convinced that *History and Class Consciousness*, in spite of its difficulties with "nature," can be a powerful ally in a struggle for a better world. Lukács matters to this book for his unique understanding of the conditions of possibility for radical change. He brings a particular understanding of the interrelationship between thought and action, making him in some senses the archetypal theorist of situated knowledges and of an immanent critique of everyday life.

Given this, the chapter is structured around three closely related conceptual debates. First, I seek to respond to Žižek's question of how we today stand in relation to the revolutionary challenges of the work.[2] This requires understanding which subject might be capable of the kind of revolutionary act Lukács defined. In certain feminist appropriations of the cyborg, I argue that we find a differentiated, fragmented subject more open to a radical environmental politics. This is a figure in the process of becoming. It suggests a form of consciousness to be struggled for and not one that is preconstituted: this disrupts both essentialist interpretations of the subject of social change and of the object of consciousness. Second, I seek to reclaim Lukács's writings for a critique of everyday life. Nevertheless, reclaiming such work involves confronting his interpretation of "imputed consciousness." This is often taken to imply that the proletariat needs assistance in grasping the perspective that is rightfully (and essentially) its own. For critics, this also implies that the Party will be required to step in as the agent that is able to impute this true consciousness into the minds of otherwise passive workers. In his brief references to this concept, Lukács appears to decouple his analysis from the concrete struggles of quotidian life. Here he valorizes one essential perspective: this, in turn, is seen by many to be that of the Party. Little could be more contrary to the position developed in this book. However, instead of following this shallow interpretation of "imputed consciousness," I seek to place far greater emphasis on the generation—and struggle for—"conditions of possibility" for radical thought and, within this, crucial *moments* of decision making. These moments, I will argue, are fundamental to the practice of a radical urban politics. This helps to resituate Lukács's work as what Stuart Hall might refer to as a marxism without guarantees and not as a latter-day justification for the totalitarian terror of Stalinism.

Finally, returning to the figure of the cyborg, I seek to go beyond Lukács's confusing, contradictory, and antinomian position on nature. Intriguingly, aspects of this latter debate have rumbled on continually since the first publication of *History and Class Consciousness*. Lukács's criticism of Engels, and in particular his rejection of the Engelsian "dialectics of nature," was to invoke the wrath of fellow members of the Hungarian Communist Party, such as Rudas, as well as of emerging "official" Soviet philosophers such as Deborin. With the discovery of Lukács's remarkable 1925/26 defense of *History and Class Consciousness* in a locked vault of the Hungarian national archive and its publication in 1999,[3] "another act in the tragedy" was played out in the words of Paul Burkett.[4] In this secret work, Lukács accepts that there is a dialectic of nature, but he refuses to extend the dialectical method to the practice of the natural sciences. In so doing, he simultaneously provides hope to and frustrates those who have sought to appropriate his work for an environmental critique. If Löwy is correct to claim that "much of the discussion about *History and Class Consciousness* has been sidetracked into a Byzantine, metaphysical dispute over the dialectics of nature,"[5] I argue that this is in part because it distracts from the more intriguing methodological insights. These insights necessarily tend toward the development of an immanent critique of everyday socio-natures.

In developing this argument, it is vital to contextualize Lukács's work and this is the task I turn to first. In spite of the monumental influence of *History and Class Consciousness*, there have been almost no references to it within the geographical literature. In part, if there is an antipathy toward Lukács on the part of geographers, this may be owing to his derogation of space. In the claim that the fragmentation of the subject leads to a contemplative stance, he goes on to argue that this "transforms the basic categories of man's immediate attitude to the world: it reduces space and time to a common denominator and degrades time to the dimension of space."[6] Smith characterizes this as "Lukács' imprisonment of space as (in turn) the jailer of false consciousness."[7] Elsewhere in the discipline, Lukács is occasionally characterized as a romantic with an antipathy to science and technology,[8] or he is remembered for his criticism of Wittfogel's *The Science of Bourgeois Society*.[9] The result of this derogation of space has been an almost complete neglect of Lukács's writings within radical geographical work. This chapter serves as one modest attempt to rectify such neglect.

Lukács's Intellectual Journey

Marshall Berman writes of Lukács as "one of the real tragic heroes of the twentieth century."[10] Thomas Mann is said to have captured him in the contradictory figure of Naphta in *The Magic Mountain*, the communist Jesuit who, as Michael Löwy has noted, illustrates the hermaphroditism of romantic anticapitalism.[11] He is best remembered for two works, *The Theory of the Novel* and *History and Class Consciousness*: it is to the latter that I will turn most attention, although it is helpful to look at the remarkable intellectual journey Lukács embarks on from *The Theory of the Novel* to his death in 1971.

In *The Theory of the Novel*, one of the signal concerns is with the dissonances that were tearing modern society apart.[12] Lukács begins with a poetic description: "Happy are those ages when the starry sky is the map of all possible paths—ages whose paths are illuminated by the light of the stars. Everything in such ages is new and yet familiar, full of adventure and yet their own. The world is wide and yet it is like a home, for the fire that burns in the soul is of the same essential nature as the stars."[13] This is the age of epic literature, as opposed to the modern era in which the novel is "the form of absolute sinfulness, as Fichte said, and it must remain the dominant form so long as the world is ruled by the same stars."[14] Whereas Lukács sees hints of a way beyond this situation in the work of Tolstoy and Dostoevsky, Löwy demonstrates how neo-Kantian dualisms lock him into a "tragic worldview."[15] It is only with a qualitative leap from neo-Kantianism to Hegelian marxism that he is able to transcend this dualism. With this *Aufhebung,* Lukács comes to argue that it is the proletariat that has the capacity to resolve the antinomies captured in *The Theory of the Novel.*

Hegel is crucial to this shift, largely because of the manner in which he transforms Lukács's reading of Marx. Löwy writes of how it was no longer through Simmel (and other associates of the Heidelberg group) that Lukács was to approach Marx, but through the German Idealist philosopher. Thus he quotes Lukács: "I no longer regarded Marx as 'an outstanding man of science,' an economist or a sociologist, but began to see in him the universal thinker and great dialectician."[16] This emerging Hegelian marxism is captured well, in the immediate aftermath of his "conversion" to marxism, in *Tactics and Ethics:* "The Marxist theory of class struggle, which in this respect is wholly derived from Hegel's conceptual system, changes the transcendent objective into an immanent one; the class struggle of the proletariat is at once the objective itself and its realization."[17] Later it reaches an even bolder proclamation in *History and Class Consciousness,* on whose opening page Lukács makes an outlandish challenge:

> Let us assume for the sake of argument that recent research had disproved once and for all every one of Marx's individual theses. Even if this were to be proved, every serious "orthodox" Marxist would still be able to accept all such modern findings without reservation and hence dismiss all of Marx's theses *in toto*—without having to renounce his orthodoxy for a single moment. Orthodox Marxism, therefore, does not imply the uncritical acceptance of the results of Marx's investigations. It is not the "belief" in this or that thesis, nor the exegesis of a "sacred" book. On the contrary, orthodoxy refers exclusively to *method.*[18]

The claim had appeared in an earlier version of the same essay, but Lukács's position changes in the intervening period.[19] Whereas the former concludes with a quote from Fichte ("one of the greatest of classical German philosophers: 'So much the worse for the facts'"),[20] in the much-revised version of the essay that appears in *History and Class Consciousness,* Löwy notes, Lukács has transformed his reading of Lenin and Trotsky (and their relation to "the facts") to one in which he criticizes a philosophical remoteness from reality *as captured* by Lenin. Here we begin to see how, although at once a culmination of his early writings, *History and Class Consciousness* is also in part an auto-critique of his own "ultra-Leftism." Ultimately this was to take Lukács on a journey to conservatism. In part, we see hints of such a movement as the book progresses; indeed, Lukács was keen to emphasize how the essays are best approached in the order in which they appear.[21] Thus, although retaining an admiration for Rosa Luxemburg throughout, in the later essays Lukács shifts his attention more to the role of the vanguard party. This sets up his next major work, *Lenin*. From 1926 onward, there is a further shift in Lukács's tone and approach: for Löwy, this is once again captured in how he reads Hegel and his relationship with marxist theory. Thus, in his 1926 essay on Moses Hess, Hegel is lauded for his "reconciliation with reality." This was later to represent Lukács's own reconciliation with the emerging regime, something Žižek characterizes as a Thermidorean moment and a betrayal of fidelity to the (revolutionary) Event itself.[22] Given these large shifts in Lukács's thought, I will restrict my focus to the period between 1919 and 1925/26— in particular to the remarkably powerful dialectical approach developed in *History and Class Consciousness*.

Dialectical Method and the Discovery of the Subject–Object of History

Throughout the social sciences and humanities, Lukács remains most closely associated with the critique of reification. It was in large part because of the proletariat's ability to go beyond the immediacy of reification and develop a dialectical understanding of the concrete totality that Lukács claimed to have found the true subject–object of history. Reification has subsequently become fundamental to a marxist theorization of ideology.[23] In part, it brings together some of Weber's concerns around formal rationality with Marx's writings on alienation and commodity fetishism. From the former, Lukács captures the encroaching scientific management of society, in which personal relationships are reduced to questions of calculation. The finished article turns into the "objective synthesis of rationalised special systems."[24] Taylorist techniques have broken down complex tasks and in the process shattered the organic unity of the labor process, thereby atomizing workers. Through developing the Hegelian roots to Marx's method, Lukács demonstrates that this makes for a subjective and an objective—an inner and an outer—transformation of society. The commodity form changes not only the objective structure of the world; it transforms people's subjective worlds. Indeed it is through the transformation of the objective world, itself dependent on human subjects, that the subjective is transformed. As Marx demonstrates, relationships between people have taken on the character of relationships between things and, increasingly, the world seems to be dominated by the things people have made. Liberation is seen not to emerge from transforming human relationships; it is seen to emerge through the acquisition of more commodities. The worker is largely unable to do this, thereby leaving her feeling utterly powerless, imprisoned in a world of things that she cannot truly possess. For the wealthy, there is the possibility of feeling far more at home in this world. Money permits wants to be provided for, even if at the heart of this there is the ideological lie that it is capital and not people providing such comfort and luxury.[25] In this regard, the commodity structure has an utterly disintegrative effect on society, remolding the latter in the image of the former. For the capitalist, money can buy love; it can make more money; and the state becomes a natural form through which society is necessarily managed. Given this, the commodity becomes "crucial for the subjugation of men's consciousness to the forms in which this reification finds expression and for their attempts to comprehend the process or to rebel against its disastrous effects and liberate themselves from the servitude to the second nature so created."[26]

This theorization of reification cannot be understood outside of Lukács's appropriation of a Hegelian method for marxist ends. Open about this intellectual debt to Hegel, Lukács writes in the 1922 preface to *History and Class Consciousness* that "we cannot do justice to the concrete historical dialectic without considering in some detail the founder of this method, Hegel, and his relation to Marx."[27] Two features define the dialectic as it is understood in *History and Class Consciousness*: first is the preoccupation with "the totality";[28] the second is a focus on subject–object relations and how the totality is first posited as an object, and second as a positing subject. Bringing these aspects together, Lukács emphasizes how this leads to an overcoming of the antinomies that beset both neo-Kantian thought and the society of which this was an integral part. The materialist dialectic must, Lukács argues, be a revolutionary dialectic, for "the central problem is to change reality."[29] The dialectical interaction of subject and object is understood here as *praxis*, and the totality is understood as a *concrete* totality: it is the product of numerous complex mediations. Thus, quoting Marx, Lukács writes: "The concrete is concrete because it is a synthesis of many particular determinants, i.e. a unity of diverse elements."[30] To grasp the concrete totality means transcending immediacy: again, this means grasping the complex mediations that produce the concrete.

"Reification and the Consciousness of the Proletariat," the most central, and arguably the most important, essay in *History and Class Consciousness*, pivots around this need to transcend immediacy. It makes the claim that bourgeois thought is simply unable to discover such mediations and therefore remains trapped within the antinomian worldview we have already encountered in *The Theory of the Novel*. On the one hand, this is because of its structural position within the production process; but, on the other hand, it is for the very reason that it would be suicidal for the bourgeoisie to attain such consciousness. Thus, whereas the bourgeoisie remain trapped within immediacy, the proletariat is able to extricate itself "by adopting its own point of view."[31] Objects of the empirical world are to be understood as aspects of a totality; facts are to be understood as processes; the state as a power factor.[32] All these seemingly immutable aspects of capitalist society are thereby denaturalized in a process-based, dialectical understanding of concrete reality.

Nevertheless, the challenge is to discover the meanings immanent to these objects, the mediations that knit the particular to the universal and make the world what it is, and also to uncover the tendencies that shape facts as processes in specific ways. These developing tendencies, Lukács claims, constitute a higher realm than reality itself. Given this, unmediated contemplation opens up a gap between the subject and object of knowledge. With the increasing dominance of the commodity form, and the transformation of objective reality into a commoditized realm, the opportunity to develop a dialectical understanding becomes more and more possible, but only for one agent—the proletariat. Because a worker's self-consciousness is also the self-consciousness of the commodity, proletarian class consciousness can achieve a fundamental change in the object of knowledge. For Lukács, the world-transforming qualities of proletarian self-consciousness are demonstrated if one is to distinguish the conditions of possibility available to the worker from those available to other subjugated groups in previous historical periods. The slave, he argues, in achieving consciousness of his or her status, does nothing to transform the object of knowledge. In contrast, the worker's self-knowledge is practical; it contributes to a transformation of the object of knowledge. In light of feminist standpoint theorists' appropriation of *History and Class Consciousness* and of Susan Buck Morss's brilliant exposition of the influence of slave rebellions on Hegel's (and presumably Lukács's) thought, this claim seems economistic lunacy. It relegates anything but anticapitalist movements to a solely peripheral status and is highly dubious. Nevertheless, if we follow the method employed, it becomes clear how Lukács could argue that a worker's self-consciousness could transform the fetishistic forms in which the commodity presents itself. In so doing, consciousness demonstrates the relational production of the concrete totality. In this way, consciousness can become social reality.

Given this, it is little surprise that *History and Class Consciousness* was to invite the criticism that has dogged it ever since from some "orthodox" quarters, that it remains an idealist work. We merely need to think differently and the world will change, so the critics claim. However, with a richer understanding of Lukács's appropriation of the dialectic, we can see how such crude claims lack foundation. Žižek captures this well, writing that if we acquire consciousness of the commodity fetish, any interruption in this necessarily involves transforming the reality in which it operates. Similarly, the critics miss the active relationship between the subject and object. As Marx claims of Feuerbach: he wants "sensuous objects, really distinct from the thought objects, but he does not conceive human activity itself as *objective* activity."[33] It is this dialectical understanding, an emphasis on the developing tendencies in society, that defines Marx's humanism from that of someone such as Carlyle.[34]

For Lukács, the gravest danger confronting society is that it might remain imprisoned within immediacy by not acquiring the dialectical understanding of the totality necessary for change (here a chink is opened through which "imputed consciousness" begins to slip in). The mechanical marxism developed by the Revisionists and finding its expression in social democracy merely increases this danger: "With the growth of social democracy this threat acquired a real political organization which artificially cancels out the mediations so laboriously won and forces the proletariat back into its immediate existence where it is merely a component of capitalist society and not *at the same time* the motor that drives it to its doom and destruction."[35] Although not altogether unproblematic again, I will argue later that many aspects of contemporary environmentalism would also seem to foreclose the conditions of possibility from which genuinely transformative socio-natural perspectives might emerge. By rooting a movement in a dualistic opposition between a benign nature and a malevolent society, the mutual codeterminations that make environments in better or worse ways are repeatedly obfuscated. The target of environmentalist critique becomes carbon or the state and not the historically and geographically situated processes through which human and nonhuman are brought together in a differentiated unity.

Achieving Standpoints, Situating Knowledges

If the group experience for workers in capitalist society generates conditions of possibility for such process-based understandings, this explains Lukács's discovery of the "messiah class." For Lukács, the proletarian standpoint *is* a privileged vantage point: "The superiority of the proletariat must lie exclusively in its ability to see society from the centre, as a coherent whole."[36] In his view, recognizing the superiority of this position is absolutely necessary for survival. For the proletariat, "to become aware of the dialectical nature of its existence is a matter of life and death."[37] By becoming object, the worker is forced to surpass immediacy. Lukács brings this to life historically by demonstrating the seminal influence of the Silesian weavers' revolt on Marx's thought. Marx revered the weavers' ability to penetrate the immediate circumstances of their oppression and thereby revolt against the *hidden* enemy, the banker: "What is crucial is how to interpret the connection between these remoter factors and the structure of the objects immediately relevant to action."[38] The standpoint of the proletariat becomes a crucial weapon in this:

> For when confronted by the overwhelming resources of knowledge, culture and routine which the bourgeoisie undoubtedly possesses and will continue to possess as long as it remains the ruling class, the only effective superiority of the proletariat, its only decisive weapon is its ability to see the social totality as a concrete historical totality; to see the reified forms as processes between men; to see the immanent meaning of history that only appears negatively in the contradictions of abstract forms, to raise its positive side to consciousness and to put it into practice.[39]

Methodologically, the starting point for feminist standpoint theorists is the same, albeit through a focus on the group experience of women in patriarchal society. Sensuous activity and phenomenological experience are shown to be crucial to the makeup of the world, the meanings that are invested in that world, and the particular understandings made possible. For Nancy Hartsock, this builds on a marxist (and Lukácsean) insistence on the epistemological and ontological priority of human activity. Thus, "the epistemological (and even ontological) significance of human activity is made clear in Marx's argument not only that persons are active but that reality itself consists of 'sensuous human activity, practice.'"[40] Hartsock claims that some feminist frustrations with historical materialist approaches lie in the latter's narrow interpretation of human activity. *History and Class Consciousness* similarly limits its focus to wage labor: the division of labor is that between worker and capitalist and not that between men and women. In contrast, by focusing attention on the gendered division of labor, standpoint theorists such as Hartsock bring a focus on "capitalist patriarchy" to the fore.[41] Beyond such a critique, Hartsock emphasizes the Lukácsean claim that the laboring subject has a privileged standpoint from which to acquire consciousness of the concrete totality. Rich potentials lie in exploring the work of women who, in societies shaped by both capitalist and patriarchal relations, are brought much closer to the material interchange with nature that Lukács and Marx saw as being so important to challenging the objective structure of the modern world.

"Given what has been said about the life activity of the proletarian, one can see that, because the sexual division of labor means that much of the work involved in reproducing labor power is done by women, and because much of the male worker's contact with nature outside the factory is mediated by women, the vision of reality which grows from the female experience is deeper and more thoroughgoing than that available to the worker."[42]

As with Lukács, there have been many criticisms of such an approach. Some find the claim to a privileged vantage point for a coherent subject to be an essentialist one.[43] Just as with proletarian consciousness, the "female experience," as Hartsock describes it, is surely likely to differ depending on one's relation to other forms of domination, whether along the lines of race, class, or sexuality. However, just as with Lukács, these criticisms fail to grasp the method through which the standpoint is understood. Hartsock's interest is not in some preconceived vantage point that simply needs to be discovered and is available only to women. Instead, she emphasizes the conditions of possibility that women's phenomenological experience of their structural position in capitalist patriarchy opens up. A feminist standpoint is *not* pregiven, it is to be struggled for and achieved through this historically and geographically situated experience. Haraway captures this well in her insistence on the political value in what she claims to be a tainted metaphor of the "standpoint" or "perspective":

> A standpoint is not an empiricist appeal to or by "the oppressed" but a cognitive, psychological, and political tool for more adequate knowledge judged by the nonessentialist, historically contingent, situated standards of strong objectivity. Such a standpoint is the always fraught but necessary fruit of the practice of oppositional and differential consciousness. A feminist standpoint is a practical technology rooted in yearning, not an abstract philosophical foundation.[44]

It is for this reason that Fredric Jameson considers feminist standpoint theorists to be deepening the "unfinished project" of *History and Class Consciousness*. Theorists such as Hartsock, Haraway,[45] Harding, and Sandoval have opened a new terrain over which to contest and transform the making of our world. They take forward the project of *History and Class Consciousness* while recognizing the partiality of its conception of the subject.

Challenging both essentialism and human exceptionalism, some standpoint feminists have turned to the figure of the cyborg, first deployed by Haraway as part of her effort "to build an ironic political myth faithful to feminism, socialism, and materialism." Since the publication of the *Cyborg Manifesto*, this figure has been put to work in numerous disparate areas to imply a commitment to a methodology that seeks to go beyond dualistic understandings of nature and society, technology and politics. For Sandoval, U.S. Third World feminists have found within cyborg consciousness "the methodology of the oppressed, a methodology that can provide the guides for survival and resistance under conditions of First World transnational cultural conditions."[46] The cyborg defies boundary policing and points to a new feminist and environmental praxis as well as an incomplete and differentiated subject in a process of becoming. In this regard, it is one important complement to standpoint approaches. Rather than contradicting the tenets on which standpoint approaches have been built, it takes standpoint approaches forward, building on the tendencies within them. In this way, it begins to speak to the question posed at the start of this chapter, as to which social agent might be able to accomplish the unique event of the early marxist Lukács. Such a figure might challenge dualistic understandings of nature and society as a way toward a progressive socio-natural praxis that defies boundaries and seeks to generate new conditions of possibility. Similarly, Kevin Floyd points to the ways in which every "standpoint" has its "blind spots": politics for Floyd is, in this sense, always aspirational as subjects seek to move beyond these blind spots.[47] I will return to some of these points later in regard to Lukács's understanding of nature/society debates. First, however, this sense in which standpoints (always considered partial) can be achieved or struggled for needs to be positioned within (or against) the problematic of imputed consciousness. For if consciousness is imputed—by a party, a theoretician, or an artistic avant-garde for that matter—it appears counter to the critique of everyday life developed throughout this book.

Imputed Consciousness or a Critique of Everyday Life?

Henri Lefebvre denigrates Lukács's theoretical outlook as a "purely speculative construction on the part of a philosopher unacquainted with the working class."[48] Here he contrasts his own reading of praxis, something formative in his development of a critique of everyday life. Paradoxically, in other contexts, Lukács's work is considered alongside Lefebvre's as crucial to the construction of a marxist theory of the everyday.[49] Lefebvre's criticisms stem, at least in part, from Lukács's apparent disdain for what workers think. Instead, he states, his interest is in what it might be possible for workers to think. Thus, given what I have argued in the preceding section, few passages could jar more than Lukács's claim that

> [such an analysis is] far removed from the naïve description of what men *in fact* thought, felt and wanted at any moment in history and from any given point in the class structure. I do not wish to deny the great importance of this, but it remains after all merely the *material* of genuine historical analysis. The relation with concrete totality and the dialectical determinants arising from it transcend pure description and yield the category of objective possibility. By relating consciousness to the whole of society it becomes possible to infer the thoughts and feelings which men would have in a particular situation if they were *able* to assess both it and the interests arising from it in their impact on immediate action and on the whole structure of society. That is to say, it would be possible to infer the thoughts and feelings appropriate to their objective situation.[50]

And "the objective theory of class consciousness is the theory of its objective possibility."[51]

In one sense, this mirrors the claim that a standpoint must be struggled for and achieved rather than simply given. However, in another sense, it appears to suggest that consciousness is preexisting and simply requires the right social agent to be able to "impute" this. Indeed, the void between actual and imputed consciousness leads Lukács's writings on class consciousness to what some would argue is a justification for Stalinism and for the Communist Party as the institutional embodiment of what proletarian consciousness should be. For John Roberts, it is the lack of a crucial mediator in *History and Class Consciousness* that leads Lukács to find a substitute in the form of the party,[52] something avoided in Lefebvre's writings through a subtle distinction between daily life, everydayness, and the everyday. Certainly Lukács's concern is much less with the actual ideas in people's minds—and here he can be both compared and contrasted with Gramsci, for whom the disorganized, contradictory common sense of the proletariat is the starting point for his own unique philosophy of praxis—but, in his emphasis on the conditions of possibility for a dialectical outlook, this does not necessarily lead to a justification of the party. Again, Lukács's criticism of those who start from what people "in fact" think must be related back to his criticism of empiricism for its failure to grasp underlying tendencies and to get beyond immediacy. In this respect, Gramsci's philosophy of praxis emerges, perhaps considerably more favorably: nevertheless, judged on these terms, everyday life remains of foundational importance for Lukács. Nowhere else can a genuinely dialectical understanding of society emerge. Thus "the continuation of that course which at least in method started to point the way beyond these limits, namely the dialectical method as the true historical method was reserved for the class which was able to discover within itself *on the basis of its life-experience* the identical subject–object, the subject of action; the 'we' of genesis: namely the proletariat."[53] And, on the relation of this life-experience to empirical reality: "The desire to leave behind the immediacy of empirical reality and its no less immediate rationalist reflection must not be allowed to become an attempt to abandon immanent (social) reality."[54] In *Soul and Form*, written prior to his marxist writings, Lukács distinguishes between authentic and everyday life, a distinction that directly influenced Heidegger's writings. This tragic (and patronizing) attitude is later transformed in *History and Class Consciousness* and *Lenin*, in which it is clear, as Löwy writes, that "there is no longer a tragic duality for the Lukács of 1924: lying beneath the surface of everyday life is a deeply rooted *tendency* toward revolution, which forms the bedrock of revolutionary activity."[55] While there is a lingering sense of essentialism in this claim of a tendency, it is better framed in terms of the conditions of possibility within the historically and geographically situated knowledges of everyday life.

This focus on conditions of possibility (as well as the claim that superseding immediacy requires direct contact with such immediacy) makes Lukács's 1919–26 writings so fundamentally important to the critique I develop in this book. Emphasizing the importance of the proletariat's struggle against reification in order to achieve self-understanding, Roberts writes: "The everyday is neither 'inauthentic' nor 'authentic,' but rather the temporal and spatial order out of which the alienations of proletarian self-knowledge will emerge."[56] Nevertheless, in line with many other critics, he goes on to castigate Lukács in a critique that distills charges of idealism (a change in consciousness changes the world) and of Lukács's notion of imputed consciousness. Similarly, for Jameson, the distinction between actual and "imputed consciousness" "opens the wedge through which Lukács's various adversaries (on the extreme left as well as on the right) glimpse the wolf-in-sheep's clothing of the Party or the Intellectual, who now conveniently get substituted for a sociological working class that needs them to find out what it 'really' thinks."[57] The problem, for someone such as Althusser, is that Lukács develops a theory of the expressive totality, in that only the proletariat—in the ideal form of the Party—can advance a genuinely marxist understanding of society. For another of Lukács's structuralist critics, Gareth Stedman Jones, "The emergence of true proletarian consciousness is tantamount to the overthrow of the bourgeoisie."[58]

I will dwell on these points for a little longer, for I think many such claims are misplaced. On the question of idealism, Lukács's own rebuttals are telling. Here we see the political tactician, wrestling with the terrain of the conjunctural. Thus, the decisive factor in the failure of the Hungarian revolution of 1918 was not, as some of his "materialist" critics such as Rudas claimed, a military failure, but the inadequate grasp on the part of the working class of the totality. Far from an idealized consciousness, only represented in the form of the Party, this is concerned with grasping the "concrete, dialectical moments of the concrete situation."[59] This implies a focus on the moment of *decision* as opposed to a "tailist-fatalist concept of process,"[60] and it is a theorization of the political conjuncture (and the relation of consciousness to this) that we find so well developed in Marx's *Eighteenth Brumaire* or in Gramsci's *Prison Notebooks*. It should be no surprise that *History and Class Consciousness*, as with those other two works, does *not* emerge as a speculative construction on the part of a philosopher, as Lefebvre claimed, but from an actor deeply engaged in and wrestling with a distinct revolutionary moment.

The "tragedy" of imputed consciousness is nonetheless somewhat harder to contend with, not least because Lukács's own justifications merely seem to reassert the criticisms made of him. This appears utterly disabling for a theory of the everyday that claims to take seriously the quotidian production of metropolitan nature as I do in this book. Thus, while claiming no particular attachment to the word "imputed," Lukács writes in *Tailism and the Dialectic* that the task of parties is to bridge the consciousness that workers "actually possess and the consciousness that they *could* have—given their class position" and elsewhere, "The form taken by the class consciousness of the proletariat is the *Party*."[61] Nevertheless, all might not be doomed.

As Feenberg has argued, Lukács's theory of the Party is actually closer to Gramsci's, in which it is clear that the Party serves as a functionary for the interests of the proletariat and not vice versa. Feenberg continues: "The party is thus not a mechanism of social control in the service of the revolution—an impossible contradiction: it is there to be 'seen,' and the sight of it is a moment in the constitution of a subject of history."[62] For Löwy, influenced above all by Goldmann, imputed consciousness "is an objective possibility . . . we can only put a wager on this subject."[63] Related to this, he picks up on Stedman Jones's complaints to argue that Lukács's emphasis is on the *condition of possibility*: imputed consciousness, in short, is a necessary condition of possibility for the taking of power, but is not the cause in itself.[64] This comes out most clearly in the case of the failed workers' republic in Hungary. Jameson too echoes such concerns, claiming that "the conception of 'conditions of possibility' then has the advantage of stressing, not the content of scientific thought, but its prerequisites, its preparatory requirements, that without which it cannot properly develop. It is a conception that includes the diagnosis of blocks and limits to knowledge (reification as what suppresses the ability to grasp totalities) as well as the enumeration of positive new features (the capacity to think in terms of process)."[65] What Jameson and Löwy both grasp is the fecundity of Lukács's conceptualization of everyday life for an immanent critique. It is within everyday life, and specifically the everyday life of the proletariat, that the conditions of possibility for a radically different world germinate. Whether or not the moment of decision making is seized is a different matter.[66] Jameson's argument is built on the back of exactly those claims made by Hartsock and others on the standpoint as something to be achieved, not given, and one that generates conditions of possibility. Nevertheless, some have criticized key feminist standpoint theorists such as Hartsock for merely substituting one essential subject for another.[67]

In a different but no less brilliant way, Žižek demonstrates how both actual and imputed consciousness are contradictory aspects of a single subject: this is the "inner self-fissure (or 'out-of–jointness') of the historical subject";[68] the authentic Leninist party "addresses the proletariat from a radically subjective, engaged position of the lack that prevents the proletarians from achieving their proper place in the social edifice."[69] This radically anti-essentialist position serves to reinstate the revolutionary content to Lukács's immanent critique of the everyday. Importantly—and clearly influenced by Luxemburg here—Lukács's emphasis is on the need to generalize the spontaneous aspects of class consciousness. Again, the Party is kicked into action by the "out-of-jointness" within the more spontaneous elements of the working class. Intriguingly, in the figure of the cyborg, this fragmented subject is seen to be a socio-natural one. For Lukács, nevertheless, the latent socio-natural critique is ultimately blocked by the very antinomies the book is targeted against.

The Critique of Positivism and the Dialectic of Nature

If *History and Class Consciousness* is considered to be one of the foundational texts of Western marxism, it is, in many ways, because the work goes against the emerging Soviet tendency of developing a scientific marxism. This tendency was to bring about an increasing rapprochement with positivist forms of knowledge production and a so-called science of the dialectic. Lukács was a fierce opponent of this mechanistic science: his reading of the dialectic was, as a result, set on a collision course with the dialecticians of the new Soviet

Academy of Sciences, of whom Deborin was quickly becoming a key representative. For Lukács, not only did positivism fragment "the totality," but it represented a nontransformative approach to theorizing reality that perpetuated a subject–object dualism. The detached observer analyzing reality is the archetype of such an approach. Dialectical thought should be utterly opposed to this: the focus on mediations surmounts such a dualism, placing emphasis on the prospects for transforming, rather than merely observing, reality.

As with Gramsci, Lukács's critique of positivism is clearest in his reaction to Bukharin's *Historical Materialism* (for Gramsci, he refers to this as the *Popular Manual*). Bukharin, Lukács holds, "instead of making a historical materialist critique of the natural sciences and their methods, i.e. revealing them as products of capitalist development, he extends these methods to the study of society without hesitation, uncritically, unhistorically and undialectically."[70] As Gramsci notes, this reduction of society to a set of laws, according to criteria set up from the natural sciences, is to deny the rich complexity of the existing world. For "the experience on which the philosophy of praxis is based cannot be schematized; it is history in all its infinite variety and multiplicity."[71] Positivism as "vulgar evolutionism is at the root of sociology, and sociology cannot know the dialectical principle with its passage from quantity to quality."[72] Lukács, similarly, contrasts dialectical thought with contemplative thought: Bukharin's "leanings towards the natural sciences and his frequently acute dialectical instinct are here inevitably in contradiction."[73]

The prelude to Lukács's critique is built on an analysis of Bukharin's technological determinism. Just as environmental determinism fetishized one particular moment of social change, so Bukharin fetishizes technological change, assuming this to bring about a new division of labor:

> The point has been stressed because it clearly reveals the essential error in Bukharin's conception of historical materialism. The closeness of Bukharin's theory to bourgeois, natural-scientific materialism derives from his use of "science" (in the French sense) as a model. In its concrete application to society and history it therefore obscures the specific feature of Marxism: that all economic or "sociological" phenomena derive from the social relations of men to one another. Emphasis on a false "objectivity" in theory leads to fetishism.[74]

In this we have the Lukács who has influenced science and technology studies—the philosopher urging us to see beyond essentialized claims in which transformative power is imputed to specific technologies and urging us to strive for the mediations that invest a subject status to technological objects and objectify social subjects.[75] From this, we can see how Robert Young might find in Lukács a solid basis for his claims around "science as culture." This dialectical conception of technology, shaped as it is by a profoundly Hegelian take on the philosophy of praxis, marks a distinctive reading of the power of specific technologies within society that potentially opens up the way to a democratization of technology. Again, this democratization is to be forged by the possible consciousness of everyday men and women who recognize their twin status as subjects and objects in an alien world. As with Young, the call is not for a better science but for a recognition that science is always bound up in power relations.[76] We should not strive for a science free of ideology but for the recognition that science will always be ideological.

It is in part also such claims that were to invite the criticism of New Left scholars such as Stedman Jones and Colletti, both of whom detect an antiscientific romanticism in Lukács's writings, alongside a subjective idealism and slippery relativism. Thus, Stedman Jones describes *History and Class Consciousness* as "the first major irruption of the romantic anti-scientific tradition of bourgeois thought into Marxist theory."[77] He continues, claiming that "there is no suggestion of the liberating effects of industrialization and scientific discovery, let alone of Marx's belief that the theory of historical materialism was itself a real and responsible science."[78]

For Feenberg, such criticisms actually confuse the issue. Indeed, elsewhere in *History and Class Consciousness*, Lukács appears perfectly happy with a contemplative approach to the natural sciences, if their scope is limited to the study of nature and not society. Here we find a marxist philosopher far harder to reconcile with a critique of technology, for it is one who is content to leave existing scientific practice free of any criticism. Thus, Lukács writes: "When the epistemological ideal of the natural sciences is applied to nature it simply furthers the progress of science. But when it is applied to society it turns out to be an ideological weapon of the bourgeoisie."[79] Simple: let scientists do what they do, free from any social criticism, but never seek to apply such schematic frameworks to a study of society. Here, through restricting science to the study of nature and dialectical materialism to society, we see a dramatic return of the very antinomies that *History and Class Consciousness* seeks to undermine. Rather than developing an entirely socialized concept of nature, as Schmidt, among others, claims of him, Lukács actually falls back on the antinomian view he criticizes.[80] This is made quite explicit in what must be Lukács's most widely cited footnote: here he clarifies his position in relation to the Engelsian Dialectics of Nature:

> It is of the first importance to realize that the method is limited here to the realms of history and society. The misunderstandings that arise from Engels' account

of dialectics can in the main be put down to the fact that Engels—following Hegel's mistaken lead—extended the method to apply also to nature. However, the crucial determinant of dialectics—the interaction of subject and object, the unity of theory and practice, the historical changes in the reality underlying the categories as the root cause of changes in thought, etc.,—are absent from our knowledge of nature.[81]

For Engels, "practice," as it was embodied in the scientific experiment, meant that the dialectical method could be extended to the natural world: in short, scientists had dissolved the Kantian "thing-in-itself" by reproducing and remaking nature in the laboratory. Again, Lukács rejects this, claiming that "Engels' deepest misunderstanding consists in his belief that the behaviour of industry and scientific experiment constitutes praxis in the dialectical, philosophical sense. In fact, scientific experiment is contemplation at its purest."[82]

The rejection of the dialectic of nature was to invoke the ire of Rudas and Deborin and was to lead to Lukács's censure at the Fifth World Congress of the Third International, at which, in his opening address, Zinoviev condemned him for theoretical revisionism that "cannot be allowed to pass without impunity."[83] The main objection to *History and Class Consciousness* was that it refused to treat the dialectic as an objective science and that this is made clear through its author's refusal to extend the dialectic to nature. Given the direction in which *History and Class Consciousness* tends, it remains remarkable that Lukács appears to restrict his analysis to "society."

Both Feenberg and Jay, nevertheless, argue that even if it exposes a serious contradiction at the heart of *History and Class Consciousness*, the restriction of his approach to society rescues Lukács from a far worse fate.[84] For Feenberg, "the positing of nature by society, its reduction to an object of social practice, is unthinkable except in the concept of a speculative philosophy of nature such as Lukács rightly rejects."[85] However, if true, this would also appear to deliver a fatal blow to our efforts to appropriate Lukács for a nondualistic approach to metropolitan nature. Although writing with only a scant knowledge of what Lukács was really arguing, Gramsci was already aware of what such a separation amounts to, while also, importantly, recognizing how Lukács's problems stem from his critique of the application of positivism to historical materialism. Thus, in the only reference to Lukács he makes in the *Prison Notebooks*, Gramsci writes:

> One must study the position of Professor Lukács towards the philosophy of praxis. It would appear that Lukács maintains that one can speak of the dialectic only for the history of men and not for nature. He might be right and he might be

wrong. If his assertion presupposes a dualism between nature and man, he is wrong because he is falling into a conception of nature proper to religion and to Graeco-Christian philosophy and also to idealism which does not in reality succeed in unifying and relating man and nature to each other except verbally. But if human history should be conceived also as the history of nature (also by means of the history of science) how can the dialectic be separated from nature? Perhaps Lukács, in reaction to the Baroque theories of the *Popular Manual*, has fallen into the opposite error, into a form of idealism.[86]

Gramsci is, I would argue, only partly correct here. Lukács does claim that there is a dialectic of nature, and it would appear that he sees human history as the history of nature; however, his critique of positivism compels him to argue for a dialectical approach that is relevant only to society. This leads to an ontological separation that is devastating to his overall argument. In a fascinating attempt to reclaim Lukács for an environmental politics, Vogel picks up on this, going against the most common criticism of *History and Class Consciousness* while building on what Feenberg and Jay see as Lukács's potentially catastrophic error:[87]

> What Colletti and Jones and Feenberg and Arato all seem to agree on . . . is that the dissolution of the boundary between the natural and the social, which is seen to follow if the critique of contemplation is applied to natural science, would represent a reduction ad absurdum fatal to Lukács's epistemology. I want rather to argue quite the contrary, that what would be fatal would be to try to graft a dualism onto the epistemology that the latter rather simply contradicts, and that Lukács's interest as a thinker lies in just the position denying the independence of nature from the social that Feenberg calls "rigorously consistent and obviously absurd.". . . This is not, I think, idealism, at least in any ordinary sense; and it is certainly not romanticism or vitalism.[88]

Burkett develops a similar argument, claiming that Lukács leaves science and technology unpoliticized and unquestioned. Thus, having noted that "the rejection of the 'dialectics of nature has long been thought of as the most fundamental factor distinguishing Western Marxism from official Soviet-style Marxism,"[89] he notes the problems in such a position. In Lukács's case, it leads to a purely external critique of natural science, rather than putting a dialectical methodology to work in developing an immanent critique of natural scientific methods. Science actually remains untouched. It is an approach relevant to the "pure immediacy" of the natural world. The mediations that shape this practice—such as class, race, and gender—are left unscrutinized.[90] Young's appropriation of Lukács is thus a misappropriation.[91] Lukács has no intention of showing that science is social relations, nor that science is culture. Science, rather, is the proper approach for the study of the natural world.

"A New Act in the Tragedy"?

Burkett developed his critique in response to the English-language publication of *Tailism and the Dialectic*, something he refers to as "a new act in the tragedy."[92] In this remarkable "secret" defense of *History and Class Consciousness*, discovered hidden in an archive in Budapest in the 1990s, Lukács seeks to clarify his critique of the dialectic of nature. For some, this clarification appears to go against the statements claiming a separation of nature and society. John Rees argues that *Tailism and the Dialectic* provides a challenge to those critics who "accused *History and Class Consciousness* of being hostile to the idea that Marx's method could be extended to account for developments in the natural world."[93] Thus in the text, Lukács claims: "Self-evidently society arose from nature. Self-evidently nature and its laws existed before society (that is to say before humans). Self-evidently the dialectic could not possibly be effective as an objective principle of development of society, if it were not already effective as a principle of development of nature before society, if it did not already objectively exist."[94] To this, Burkett responds by arguing that, whereas Lukács may well recognize a dialectic in nature, this does not lead to him extending the dialectical method to a study of the natural world.

Tailism and the Dialectic does indeed support the claim that Lukács saw nature as socially conditioned:

> "For what my critics call my agnosticism is nothing other than my denial that there is a socially unmediated, i.e. an immediate relationship of humans to nature in the present stage of social development. . . . Therefore, I am of the opinion that our knowledge of nature is socially mediated; and so I remain true to the Marxian formulation of the method of historical materialism: 'it is social being that determines consciousness.' How a dualism (a dualism of society and nature) is supposed to arise out of this conception is unfathomable to me."[95]

Here we find further evidence for Vogel's claims, even if the impact of his argument is lessened by the fact that Lukács has already argued it.[96] This also backs up the often-criticized claim in *History and Class Consciousness* that "nature is a societal category. That is to say, whatever is held to be natural at any given stage of social development, however, this nature is related to man and whatever form his involvement with it takes, i.e. nature's form, its content, its range and its objectivity are all socially conditioned."[97] However, it also reinforces Lukács's somewhat naive references to the role of science. Thus, even if *Tailism and the Dialectic* does not provide a resolution to our difficulties, it does add texture to an understanding of Lukács's approach to science and to nature. The position he seems to arrive at is that, on the one hand, nature is socially mediated; on the other hand, the dialectical method should not be extended to the natural world.

In one final twist, as with so many of Lukács's tortuous auto-critiques of his own work, the 1967 preface to *History and Class Consciousness* throws these matters into confusion all over again. On the one hand, Lukács attacks his own claims to nature as a social category while also claiming that he misunderstood this point because of an inadequate recognition of the role of labor in mediating the metabolic interaction of society and nature: "The purview of economics is narrowed down because its basic Marxist category, labour as the mediator of the metabolic interaction of society and nature, is missing." And, in making such an error, he claims that "it is self-evident that this means the disappearance of the ontological objectivity of nature."[98] Needless to say, Lukács's critics have picked up on this claim as a late-in-the-day rejection of the claim in the original work that nature is a social category.[99] Once again, our efforts to put Lukács to work in a nondualistic critique of everyday socio-natures founder, through his stubborn insistence, against the overall movement of his argument, on the "ontological objectivity of nature." Nevertheless, as Vogel notes, here perhaps we also see a recognition that the whole mess results from an inadequate, idealist conception of the role of labor within the making of natures. Developing Vogel's claim, I wish to argue that many of Lukács's difficulties may well be overcome through a conception of the production of nature that places far greater emphasis on the practical activity of human and nonhuman out of which socio-natures are constituted. This is exactly the position I have developed so far in this book.

Working with and beyond *History and Class Consciousness*

In the course of this chapter, I have argued that we need to go further than *History and Class Consciousness* in several key areas. First, we need to reinterpret the revolutionary messiah class, so important to Lukács's original work, and question who is capable of achieving the radical transformation of society necessary to remake reality. Reading this subject as a struggling, dislocated, and not-quite-constituted agent is vital for undermining any tendency toward essentialism. Second, the crucial moment of decision making needs to be drawn out to undermine any sense of teleology. Conscious subjects are forced to make decisions under conditions not of their own choosing. These decisions alter the course of history, as Lukács himself was perfectly aware following his active involvement in the Hungarian Revolution. Finally, these decisions themselves and the revolutionary humanism of *History and Class Consciousness* need to be subjected to a socio-natural critique. Even if Lukács is seen to hove to an understanding of nature as socially produced, this claim merely metes out a dualism of society and nature that is related externally: this is not a dialectical view of socio-natures as something ontologically consistent and woven through human and nonhuman practice. Here, once again, I think Haraway has it partly right: "What nature could not be in these Marxist humanisms is a social partner, a social agent with a history, a conversant in a discourse where all of the actors are not 'us.'"[100] Nevertheless, I remain convinced that nature *can* be a partner in the overall framework Lukács provided, just as I believe that we need not fall prey to essentialisms or teleology.

In this regard, *History and Class Consciousness* remains a unique resource for developing an immanent critique of everyday socio-natures. Indeed, it constitutes perhaps the most theoretically sophisticated foundation within a marxist tradition for countering the dualisms that have plagued environmental thought. Developing the dialectical method through a distinctive philosophy of praxis orientates the book toward a process-based understanding of the world. Here subject and object develop in a process of mutual coevolution. Sensuous human and nonhuman entanglements make reality. These sensuous entanglements are historically and geographically specific, such that in Western capitalist societies the relationship is one in which processes appear as atomized things. Even if Lukács ultimately fails in extending his nondualistic argument to science, technology, and nature, the overall tendency of his argument is to view the world as a complex differentiated unity. If this much is present within Marx's own writings, nowhere is it sufficiently developed. From the sustained argument for the dialectical method as the defining feature of "orthodox marxism" to the nuanced consideration of the antinomies of bourgeois thought, *History and Class Consciousness* remains a uniquely powerful resource for explaining how dualisms arise in both everyday life and philosophical thought, as well as what forms of praxis might counter them.

This argument develops as part of an immanent critique of the everyday. The pivotal argument to *History and Class Consciousness* is that everyday men and women have the ability to become conscious of how the world is constituted. Through developing such a transformative consciousness, Lukács demonstrates the potential for revolutionary subjectivities within the quotidian. This, as I will argue below, is a lesson crucial for contemporary environmentalism and crucial for a democratic, transformative politics. As both Žižek and Jameson have shown in different ways, this unfinished project can be taken forward by reconsidering the collective revolutionary subject in nonessentialist ways. Feminist standpoint theory has made a fundamentally important contribution here. Finally, in relation to this, *History and Class Consciousness* becomes a powerful argument for the importance of situated contextual knowledges. Situated knowledges are concerned with the mediations that compose the socio-natural world: these knowledges are to be struggled for as a way of transforming the world. If the direction of these arguments is to be world changing, they also need to be reconsidered and reworked if they are further to develop an immanent critique of the *nature* of everyday life. In this final section, I will attempt to do this. So far, I have sought to ground each theorist within an understanding of the socio-natural production of the contemporary city.

For Swyngedouw, the mutual coevolution of socio-natures and technological things can be understood as a form of cyborg urbanization,[101] or, as White and Wilbert term it, the making of "urban technonatures." Each of these metaphors seeks to capture the way in which the city is an emerging socio-natural network combining circulations and metabolisms of human and nonhuman. For White and Wilbert, the trope of technonatures is not intended to capture an epochal shift in the constitution of the world: rather, it is employed as a hopeful metaphor to counter claims of the death of environmentalism.[102] Building on Haraway's work, they claim that "there is no sense of nature, human subjectivity, the body, or even concepts of sustainability and ecology that can effectively be thought outside of, or separable from, ever more technologized societies and social relations." They seek to build on transformative possibilities and thereby move beyond the dualisms and oppositions of earlier environmentalisms. Here "there is a structure of feeling in these 'cyborg ecologies' that there is now no going back to any kind of purism of the natural."[103] Harvey too captures this in different ways when he claims that there is nothing particularly unnatural about New York City.[104]

Much of contemporary environmentalism, just as with social democracy in the 1920s, does not strive for the mediations through which the cyborg city is constituted as a socio-natural assemblage. Instead, the social and the natural are separated through the desire to preserve green space, show the impact of humans on nature, or to show nature's revenge on an unthinking human populace. As I have already argued, this hobbles the contemporary movement and fosters an elitism and antidemocratic tendency that is not only disabling but is politically dangerous. *History and Class Consciousness* provides us with a deeply sophisticated theoretical framework for moving beyond any such dualisms, allowing us to capture these urban technonatures or cyborg cities. It does so, as I have argued, through its appropriation of the dialectical method underlying Marx's philosophy of praxis. Reality, Lukács argues, is a complex of mediations in the process of becoming. Regardless of class position—and we might assume gender, race, and sexuality—each one of us lives in a state of immediacy: those most closely involved in the metabolic interchange that makes up production are able to penetrate this immediacy and to begin to grasp, and thereby transform, the mediations out of which reality is constituted. In this regard, Lukács's philosophy of praxis takes us beyond many of the heuristic frameworks through which researchers have begun to interpret urban technonatures. Whereas actor-network theory disrupts dualisms and takes us in bold directions toward interpreting (but never changing) networks of human and nonhuman actants, the very metaphor of the network fails to capture the way in which human and nonhuman, person and thing, technologies and "nature," penetrate one another, utterly transforming their content and form. Through the dialectical method Lukács puts to work, this is exactly what he seeks to explain. This, he argues, is at the core of what distinguishes Marx's humanism from that of others. The human and nonhuman do not merely relate to one another; they utterly transform each other. Whereas "the bourgeoisie perceives the subject and the object of the historical process in a double form,"[105] this rigidly twofold form fails to reflect "the dialectical character of the historical process in which the mediated character of every factor receives the imprint of truth and authentic objectivity only in the mediated totality."[106] This mediated totality is constituted through sensuous human practice.

Clearly, such a framework cannot quite capture the direction in which Haraway's work, in particular, has moved. For Haraway, the conception of practice remains a problematic one, implying a human exceptionalism on which the entire project of marxist humanism founders. Indeed the human, as a category existing prior to its socio-natural constitution, is continually disrupted within Haraway's work. Nevertheless, perhaps *History and Class Consciousness* can be read as a related venture. If, as Haraway terms it, "'becoming with' is a process of becoming worldly," then there is the possibility of reading Lukács's critique of capitalist society in the same way. Reality is *not*, Lukács argues, it becomes: this becoming of reality, I would argue, must be understood as one that combines human and nonhuman, technical and nontechnical, in historically and geographically situated, mutually constitutive ways.

The immediacy of the contemporary city, nevertheless, is one in which things and technologies have acquired a power over their human others. The city, in its immediacy, is the antithesis of the natural. This is the argument of

Kaika and Swyngedouw, who demonstrate how the opacity of urban infrastructural networks serves to fetishize the modern city.[107] Rather than acquiring a consciousness of the socio-natural mediations through which the city is constituted, most of us live within this reified immediacy, or rather within the "phantasmagoria of urban technological networks." Such claims lead Kaika to a demonstration of how, in moments of crisis, the fetishism of the urban infrastructural ideal is disrupted.[108] Nevertheless, for Lukács, it is not only in moments of crisis that the mediations constituting the socio-natural are revealed. While this may be the case for the bourgeoisie (for whom the rigidity of the subject–object dualism "can only be broken by catastrophe"),[109] it is not so for the proletariat. For the worker, it is a matter of life and death to go beyond this immediacy and, through the class struggle, the objectification of the qualitative aspects of a worker's life can be seen as stark reality. The worker is forced to objectify him or herself as a commodity: the worker's "immediate existence integrates him as a pure, naked object into the production process."[110] Becoming conscious of this object status introduces self-consciousness to the commodity structure, which brings about an objective change:

> The specific nature of this kind of commodity had consisted in the fact that beneath the cloak of the thing lay a relation between men, that beneath the quantifying crust there was a qualitative, living core. Now that this core is revealed it becomes possible to recognize the fetish character *of every commodity* based on the commodity character of labour power: in every case we find its core the relationship between men, entering into the evolution of society.[111]

This "relationship between men" clearly needs to be conceived of as a far more complex web of relationships between a much wider range of human and nonhuman agents. One crucial aspect neglected by such a framework, as Hartsock demonstrates, is the gendered division of labor. Oddly, this gender-blind approach has been taken forward within much of the writing on urban technonatures and the cyborg city. In the case of the latter, the omission is surprising, given that Haraway's original *Cyborg Manifesto* was "an effort to build an ironic political myth faithful to feminism, socialism, and materialism." Furthermore, as Castells was to note over three decades ago, struggles over urban infrastructure are located at the point of social reproduction and not necessarily at the point of production. They are struggles over collective consumption that, more often than not, and especially in the global South, affect the lives of women more directly than men. For anthropologists such as Claude Meillassoux, unwaged work around this kind of social reproduction serves as a form of primitive accumulation or indirect imperialism.[112] Low wages are made possible through the "invisible" work of social reproduction. In this regard, production and reproduction are not located in separate "spheres"; they are woven out of overlapping socio-natural relationships. Again, this is the crux of Harold Wolpe's argument: apartheid in South Africa emerged as a response to growing militancy owing to the low wages of a racialized capitalism relying on the ever-diminishing opportunities to exploit unwaged social reproduction in the Reserves.[113]

This situation continues in many cities today. Within my own research in Durban, South Africa, I focused on the development of a free basic water policy under conditions of increasing fiscal austerity. The municipal water-services provider in the city introduced what, on the surface, appeared a deeply progressive policy of extending access to 6kl of free basic water to all within the city. Nevertheless, with increasing pressure coming from the city's bulk-water provider to ensure a return on its infrastructural investments, this minimum allowance quickly became a maximum allowance for many households. The responsibility for ensuring the survival of the household under these conditions has, increasingly, fallen on women. Although the majority of reproductive labor within poorer households is not sold as a commodity, this work is positioned more profoundly than ever at exactly the point of dialectical antithesis of quality and quantity. Seeking to provide sufficient water for the survival of themselves and their families, many poor women confront most directly what Swyngedouw refers to as the "transformation of local waters into global money."[114] Just as Marx recognized in the conscious acts of the Silesian weavers the ability to go beyond the immediate situation and to confront the hidden enemy, we see women struggling to transform the socio-natural relations through which clean drinking water is produced and exchanged as a fiercely rationed commodity. Consciousness of the mediations through which this cyborg city is produced is far from pregiven: rather, it must be struggled for through contextually rooted and relationally defined "situated knowledges."

The development of such situated knowledges in such circumstances speaks to the point I raised earlier as to who might be able to get to grips with such socio-natural transformations. If the labor that mediates the metabolic process of reproducing the household is recognized to be predominantly that of women, this is to argue that urban technonatures are deeply gendered. Those involved in producing these networks and mediating the complex entanglements of socio-natural technologies are most frequently women. Again, building on Žižek, this is not a claim addressed to some preexisting universal identity—

Mohanty's "militant third world woman." Here, "the 'universal mission' of the proletariat arises from the way the proletariat's very particular existence is 'barred,' hindered from the way the proletariat is a priori ('in its very notion,' to put it in Hegelese) not able to realise its very *particular* social identity."[115] This approaches Haraway's attempt, quoted earlier, to rescue the standpoint from its tainted metaphor, an attempt deeply inspired by Sandoval's call for oppositional consciousness as the basis for a feminist project.[116] Herein lie the conditions of possibility for such oppositional consciousness to remake reality in fundamental ways.

Finally, this must be seen as an understanding rooted in the everyday. Appropriating Lukács for our own purposes therefore requires recognizing that socio-natures are produced and reproduced through sensuous entanglements in everyday places. White and Wilbert touch on this in their review of emerging work, pointing to "a range of brilliantly imaginative technonatural geographies and sociologies of everyday life (e.g. Michael, 2000; Clark, 2002) that seek to capture the phenomenological experience of how natures and modern built environments have become increasingly interwoven in the everyday experience of diverse peoples in such things as garden filled atrium buildings and malls, tourist resort developments, enhancements of ecological reserves, or in serious leisure activities like walking."[117] Having already stated that the distinction between the human and the nonhuman ceases to have the meaning that was once conferred upon it (as we recognize our own multiple entanglements with others), this statement appears platitudinous: of course this would be the case for everyday life if it is the case for anything at all. What Lukács pushes us toward is to recognize how "in this phenomenological experience" there is the possibility for a radical critique.[118] At the heart of such an immanent critique is, I would argue, the very real possibility for producing our cities as fundamentally more progressive socio-natural formations. This has to be at the center of a genuinely democratic environmentalism fit for purpose. Struggling for a dialectical, oppositional consciousness of how our cities are produced within and through everyday practices is a fundamental first step toward changing them for the better.

Conclusions

In this chapter, I have turned to the ways in which the sensuous entanglements out of which our cities are made come to exert a power over those life forms within them. If cities are complex socio-natural assemblages, they are also places in which, under certain historically and geographically specific conditions, some technologies extinguish the full possibilities for living a happy life. In Durban, "trickler" technologies limit people's access to water, regulating the rhythms of daily life through dictating when people can access water and when they cannot. In a rich critique of the apolitical stance adopted by actor-network theorists—to whom such nonhuman agency would be of little surprise (see, for example, Marianne de Laet and Annemarie Mol's work on the Zimbabwe bush pump)[119]—Scott Kirsch and Don Mitchell make the compelling point that it is perfectly obvious that nonhuman agency exists, but the point should be to explain this.[120] If technologies regulate day-to-day existence, rather than marveling at it, perhaps we need to explain and transform this situation. This chapter has been an attempt to get to grips with the processes and relationships that make cities in this way. In so doing, I have turned to Lukács's *History and Class Consciousness*.

Few works of marxist theory have extended the Hegelian method as far as Lukács does. Showing that the defining feature of orthodox marxism is the dialectical method, he then takes this forward in a radical attack on antinomian thought. If philosophical thought is hobbled by subject–object and nature–society dualisms, this is in large part because it has been unable to grasp the interpenetration and mutual coevolution of subject and object. With the rise of the commodity structure in capitalist societies, human life is objectified through the production process. Workers become a series of inputs and outputs, to be laid off when times are lean and to be bought when times are good. Of course, commodities, capitalist society, and indeed reality would not exist were it not for the laboring acts of such objects. In this way, workers are the subject–object for whom it might be possible to transform the commodity structure and therefore transform reality.

At the heart of Lukács's work is a nondualistic approach to understanding the world. Building on Marx's philosophy of praxis, the antinomies of bourgeois thought are seen to be resolved through praxis. Nevertheless, Lukács's vehement opposition to the application of scientific methods for the study of human societies led him to separate society and nature. For some, this saves him from the folly of dissolving nature into society. Nevertheless, for this book, this is a catastrophic error on the part of Lukács. He undermines his own argument and fails to confront reality as it actually is—a complex of socio-natural processes that refuses any discrete categorization. This is the fundamental lesson of posthumanisms, but it is also a lesson of Smith's "production of nature" thesis, which I argue can be radicalized through conversations with a Lukácsean framework.

Thus, building on what I argue is the direction, or, better still, the method of Lukács's approach (here we might paraphrase his own words on Marx:

Orthodox Lukácseanism does not imply the uncritical acceptance of Lukács's investigations. . . . On the contrary orthodoxy refers exclusively to method) leads us toward a socio-natural conception of reality that may well be appropriated for an immanent critique of the nature of everyday life. Necessarily, this requires a nonessentialist understanding of who may be able to struggle for the dialectical consciousness of metropolitan nature that is adequate to the task of changing the world. And necessarily, this means recognizing that the moment is one to be seized by conscious agents and not one that blindly unfolds. With this, I would argue, we have the possibility for thinking through conditions of possibility for democratically remaking our cities. How these conditions of possibility articulate with preexisting histories and geographies is the concern of the next chapter. Here I turn to the insights of Antonio Gramsci and the manner in which they develop over the terrain of the conjunctural.

Chapter 4 When Theory Becomes a Material Force
Gramsci's Conjunctural Natures

> What is philosophy? In what sense can a conception of the world be called a philosophy? How has philosophy been conceived hitherto? Does the philosophy of praxis renew this conception? What is meant by a "speculative" philosophy? Would the philosophy of praxis ever be able to have a speculative form? What are the relationships between ideologies, conceptions of the world and philosophies? What is or should be the relationship between theory and practice? How do traditional philosophies conceive of this relationship? etc. The answer to these and other questions constitutes the "theory" of the philosophy of praxis.
>
> —Antonio Gramsci, *Selections from the Prison Notebooks*

> And so I think you should stay here with your sword drawn if you're set on it and your anger is big enough. You have good cause I admit. But if your anger is a short one, you'd better go.
>
> —Mother Courage, in Bertolt Brecht, *Mother Courage and Her Children*

TO CONSIDER THE IMMANENT CRITIQUE discussed in chapter 3 as abstract from the historically and geographically situated practices that permit it—as some vain metaphysical fancy—would be to render it utterly powerless. Instead, we need a clearer understanding of the mutually constitutive ways in which theory and practice shape each other in definite historical and geographical contexts. Returning to South Africa, the small protest that arose in Amaoti in February 2003 built on a history of intense struggle that is etched into the built environment of the settlement. In turn, the laboring acts of provisioning a household with water in Amaoti have been deeply influenced and shaped by such struggles. Laboring bodies and militant subjectivities have been shaped in relation to the racialized categories of apartheid, a gendered division of labor, and the postapartheid economic order. Understanding the development of these practices and the mutually constitutive relationship they have with militant possibilities calls for a return to the "philosophy of praxis." No theorist is as closely associated with this peculiar phrase as Antonio Gramsci. The core of his *Prison Notebooks* was an attempt to take forward marxism as a historically and geographically situated immanent critique.[1] Thus, it is one of the signal contributions of Antonio Gramsci's writing to focus our attention on this earthliness of thought and its vital role in permitting and constraining the transformation of theory into a material, world-changing force. In this chapter, I will seek to follow Gramsci's lead and focus on the historically and geographically specific terrain over which militant knowledges and world-changing practices have coevolved in one South African informal settlement.

In doing this, I want to dwell for a little longer on what happens to sparks of radical insight when they appear in the world. To take Holloway's metaphor of the hidden volcano that resides within each of us,[2] I wish to question the ways in which the anger he describes can be transformed into the kind of slow-burning rage that Mother Courage demands of us (before she goes on to capitulate herself). It was of course to such questions that Gramsci devoted significant parts of the *Prison Notebooks*. The great weight he afforded subaltern perspectives was balanced with a recognition that these frequently appear as scattered and disorganized fragments. The notebooks were, at least in part, an effort to understand how these sometimes contradictory insights might be organized and the role to be played by a reinvigorated philosophy of praxis in transforming theory into a material force.

In large part because of this new theorization of the relationship between theory and practice, Gramsci is quite rightly recognized as one of the richest theorists of subaltern practices and thought. His provocative claim that all are intellectuals is an explicit recognition of the potential for critical insights to emerge within quotidian life.[3] Gramsci goes on to argue that it is "essential to destroy the widespread prejudice that philosophy is a strange and difficult thing just because it is the specific intellectual activity of a particular category of specialists or of professional and systematic philosophers."[4] This is richly suggestive for our attempts to develop a radically democratic politics from the day-to-day making of natures. Critical insights, moreover, are to be found in the practical experience of changing the world. A future society is to be born from the historical knowledge of past efforts at making reality. If we take Smith's claims from chapter 1 seriously,[5] this practical experience is necessarily one that involves the production and reproduction of nature in historically and geographically specific ways. Nevertheless, Gramsci is keen to distinguish implicit conceptions of the world from explicit conceptions: "It follows that the majority of mankind are philosophers in so far as they engage in practical activity and in their practical activity (or in their guiding lines of conduct) there is implicitly contained a conception of the world, a philosophy."[6] Here, as elsewhere, Gramsci stresses the unity of theory and practice for which he is quite rightly so well known: within the passage, however, there is also the recognition that no automatic leap is possible from implicit knowledges—suggested in practice—and an explicit theorization of the world that could capture the imagination of broad swaths of people. This is one of the crucial tasks Gramsci poses for the development of a philosophy of praxis.

Gramsci's writings demonstrate an acute awareness of the possibility for everyday, radical critique emerging through the practical acts of subaltern groups. Having seen the revolutionary optimism of the early 1920s dashed by the Fascist seizure of power in Italy, Gramsci was nevertheless aware of the need for this radical critique to take seriously the obstacles ahead of it. Developing this critique, he turned his attention to the manner in which worldviews came to cohere for specific groups within specific places and how a revolutionary force might turn this to its own advantage. Following both Marx's and Engels's recognition that the theoretical critique emerging from the proletariat can potentially become a material, world-changing force, Gramsci concerned himself with the organic and inorganic relationships between intellectuals and everyday people, between the party and the masses, between spontaneous ideas and longer processes of education and movement building. In some ways, the *Prison Notebooks* provides an important corrective to more recent autonomist writings in which any form of radical critique has been simplistically celebrated for its very superficiality and ephemerality.[7] (As Mother Courage knew only too well, "The screamers don't scream long, only half an hour, after which they have to be sung to sleep.")[8] Nevertheless, this stick can be bent too far in the other direction: here, Gramsci's philosophy of praxis is read as that of a moralizer, obsessed with the educative process. Gramsci is seen to fall back on the Communist Party as the true agent of change, an agent needed for forging a strict, tightly organized counter-hegemony. This appears to be the criticism of Richard Day, among others.[9] In several respects then, I want to keep these ideas in tension in this chapter. There is both openness to radical change in Gramsci's writing and an awareness of the lengthy transformative process ahead. Often captured through military analogies of a war of position and a war of movement, the task of achieving fundamental social change is colossal. But it is also realizable.

In this chapter, then, I am more directly concerned with how the radical ideas of subaltern peoples come to cohere within and between specific historical and geographical contexts. In doing this, I will begin with Gramsci's understanding of nature. If implicit conceptions of the world are intimately bound up with people's sensuous interactions with nature, the environment is a key terrain over which they might be consolidated and contested. The production of nature, as it has emerged through previous chapters, is a process through which both alienation *and* conditions of possibility are generated. I will argue that it is also one in which hegemony is consolidated and contested. To see this process in a one-way manner would, nevertheless, be naive and simplistic. Instead, it becomes important to understand how ideas themselves inform the production of nature and how they are woven through, and come to inform, new conditions of possibility. As with Gramsci's own writings, this requires sensitivity to the historical and geographical specificities within which these ideas cohere. It requires attention to the terrain of the conjunctural. Thus this chapter acquires a radically different shape from others in the book. It places far more weight on the historically and geographically situated practices making up the informal settlement of Inanda. Here it becomes possible to wrestle with Gramsci's ideas in a far more concrete setting. Theory, though crucial, acquires its force in real-world situations.

Inanda, as will become clear, has often been something of a barometer for broader changes taking place within South Africa. During the apartheid years, pockets of interracial harmony were free to develop owing to the peculiarities of landownership in the area. At the same time, the apartheid government was able to exploit hostility toward a predominantly Indian landlord class in an effort to reinstate the separate development at the heart of its racist ambitions. These processes were fought through and inscribed in the built environment of the settlement: new forms of housing and service delivery were piloted that generally reflected the power of the state and capital. Nevertheless, the settlement harbored some of the most vibrant and active civic movements in the late years of apartheid: these movements fought hard to rework the political ecology of the settlement in radically democratic ways. Charting the interrelations between the socio-natural fabric of Inanda, the laboring acts of its men and women, the successes and failures of social movements, and the ideas that occasionally cohere into dramatic world-changing forces provides a crucially important insight into the nature of everyday life and the potentials for radical change. Gramsci remains an unparalleled companion on this journey.

Gramsci's "Orthodox" Philosophy of Praxis

In some ways echoing Lukács's call for an "orthodox marxism" that "refers exclusively to method," Gramsci writes, in one of his scattered notes:

> Orthodoxy is not to be looked for in this or that adherent of the philosophy of praxis, or in this or that tendency connected with currents extraneous to the original doctrine, but in the fundamental concept that the philosophy of praxis is "sufficient unto itself," that it contains in itself all the fundamental elements needed to construct a total and integral conception of the world, a total philosophy and theory of natural science, and not only that but everything that is needed to give life to an integral practical organisation of society, that is, to become a total integral civilisation.[10]

The irony, of course, in these theorists' orthodoxies is that both are highly *un*orthodox in their attempts to liberate historical materialism from an adherence to sacred scripts. Their mutual claims to orthodoxy are constructed around a set of principles through which a philosophy of praxis might be built. These generative principles form the basis of a methodology for considering the radical transformation of society. These principles provide, in the words of Stuart Hall, a marxism without guarantees.[11] For Peter Thomas, Gramsci's philosophy of praxis can largely be distilled into three related methodological frames: an absolute historicism, absolute immanence, and absolute humanism. Absolute historicism refers to the specific relationship between thought and action in different historical moments. This is defined against the teleological, idealist, and metaphysical conceptions of historicism found in the work of some of Gramsci's predecessors and contemporaries. Revolutionary change must be understood to emerge through people's acquisition of consciousness of the concrete conditions in which they act. This relates to the second term, absolute immanence. For Gramsci, a revolutionary "coherent" conception of the world is gained not through ideas that are transplanted from the educated to the noneducated. Rather, this is through the efforts of everyday men and women to grapple with existing reality, something Gramsci termed an "absolute earthliness of thought." "Coherence" here is a translation of the Italian *conseguente* and when used by Gramsci implies not a *logical* coherence (and the patronizing, didactic connotations that would come with this) but a fusion of theory and practice. The final term, "absolute humanism," would appear to set Gramsci up for criticism from a variety of quarters, especially within contemporary (post-) human geography. Nevertheless, as Thomas shows, Gramsci's absolute humanism considers the historically varying concept of human nature and how it can only ever emerge through the coevolution of nature and society within a specific set of historical and geographical circumstances.

Returning to the passage quoted above on orthodoxy, Gramsci provides us with a sense of where some of his fundamental interests lie. First, he is interested in the construction of a "total and integral conception of the world." This involves the transformation of implicit conceptions into explicit, "coherent" conceptions that can be verbalized. Second, such a philosophy of praxis will encompass philosophy and a theory of science. And further, it will inform action and the construction of a future civilization. Gramsci's ambitions for a philosophy of praxis are not small.

In spite of these ambitions and the importance Gramsci appears to give to the philosophy of praxis, the term has often been taken as a codeword for marxism to evade the watchful eye of the prison censor.[12] Against this, Maurice Finocchiaro claims that the philosophy of praxis "points to a distinctive theoretical position,"[13] one that comprises a set of distinct concerns: "respectively, secular religion, Croce's metaphilosophy, Marx and science."[14] Wolfgang Fritz Haug develops a more convincing case, and relates this more directly to Gramsci's engagement with the *Theses on Feuerbach*. Arguing that Gramsci's use of the term from the seventh notebook onward is indicative not so much of the increased censorship of the prison environment but of a more conscious development of a theoretical position, he writes:

> One can view the unfinished Gramscian project as the new founding of a Marxist philosophy.... In order to conceive of an open becoming, Gramsci therefore needed a concept which takes up the historical figures of Marx and Marxism, but at the same time sees history after Marx, the future in the past and the horizon of the coming possibilities in a categorically open-ended fashion. This thinking is "philosophy" in a sense that Marx implicitly presupposed and only perfunctorily explained in passing and which hardly occurred to him to conceive of as philosophy. *It would be a coherent, but non-systemic thinking which grasps the world through human activity*.... It is a thinking that indeed addresses the whole, but from below with a patient attention to particularity.[15]

For these reasons, Haug argues the philosophy of praxis is "the core concept for the Gramscian project."[16] Gramsci's project thereby goes well beyond being a distinctive epistemological position, developing as it does an ontological claim, while also collapsing the separation between ontology and epistemology on which previous "philosophies" had been based. Human action and human practices come to make reality, something we have seen throughout the preceding chapters. Gramsci dwells on a phrase that he (incorrectly) ascribes to Marx, stating that "the German proletariat is the heir of classical German philosophy."[17] This statement seems to capture the move that Marx makes, as the idealist philosophy of Hegel is transformed into a theory of revolutionary social change in which everyday men and women are able to make history. Alienation finds its dissolution through overcoming the relations of capitalist society and not through a mental act. Gramsci draws sustenance from the *creative* element that he so admires within classical German philosophy while seeking a new unity of theory and practice. Throughout, we find an emphasis on this unity of theory and practice, of understanding the world through changing it, and of the "active-man-in-the-mass" as the starting point for a critical praxis. This creative, critical praxis finds its expression through what, in particularly striking terms, Gramsci refers to as "the cathartic moment":

> The term "catharsis" can be employed to indicate the passage from the purely economic (or egoistic-passional) to the ethico-political moment, that is the superior elaboration of the structure into superstructure in the minds of men. This also means the passage from "objective to subjective" and from "necessity to freedom." Structure ceases to be an external force which crushes man, assimilates him to itself and makes him passive; and is transformed into a means of freedom, an instrument to create a new ethico-political form and a source of new initiatives. To establish the "cathartic" moment becomes therefore, it seems to me, the starting point for all the philosophy of praxis, and the cathartic process coincides with the chain of syntheses which have resulted from the evolution of the dialectic.[18]

Whereas practical activity can be the basis for a philosophy of praxis, as Haug demonstrates,[19] this always articulates with previous historical experiences and with a range of sedimented knowledges. Indeed, these knowledges serve as the terrain over which a total and integral conception of the world may or may not "cohere." Thus, there can be no guarantees that experience will be translated into a radical philosophy. In part, Gramsci considers this possibility through the concepts of "common sense" and "good sense." The former refers to the almost unconscious thought taken from the past, often disorganized and lacking coherence. As Stuart Hall writes, common sense "is the terrain of conceptions and categories on which the practical consciousness of the masses of the people is actually formed. It is the already formed and 'taken for granted' terrain on which more coherent ideologies and philosophies must contend for mastery; the ground which new conceptions of the world must take into account, contest and transform."[20] Acutely aware of the material force these conceptions of the world are able to acquire and, presumably building on Marx's *Introduction to a Critique of Hegel's Philosophy of Right*, Gramsci writes:

> Marx is referring not to the validity of the content of these beliefs but rather to their formal solidity and to the consequent imperative character they have when they produce norms of conduct. There is, further, implicit in these references, an assertion of the necessity for new popular beliefs, that is to say a new common sense and with it a new culture and a new philosophy which will be rooted in the popular consciousness with the same solidity and imperative quality as traditional beliefs.[21]

Good sense therefore refers to the emergence of critical thought through achieving a unity of theory and practice. It refers to the effort to assemble the "rough and jagged" beginnings of a new world, represented in embryonic, implicit conceptions of the world, into a "coherent" but, as Thomas and Haug both make clear, nonsystemic, worldview that remains in an organic relationship with those who have produced it.[22] Philosophy, in short, coincides with "good sense."[23] Whereas Gramsci is well known for placing great emphasis on the role of education and for opposing aspects of what he saw as spontaneism within the writings of Rosa Luxemburg,[24] it is important to recognize that Gramsci's position is not a reformist one. Nor is it a crude argument for the lifting of a veil of false consciousness from the eyes of the masses. As Gramsci writes of the (male) worker: "His theoretical consciousness can indeed be historically in opposition to his activity. One might almost say that he has two theoretical consciousnesses (or one contradictory consciousness): one which is implicit in his activity and which in reality unites him with all his fellow workers in the practical transformation of the real world; and one superficially explicit or verbal, which he has inherited from the past and uncritically absorbed."[25] Clearly, good sense is available to anyone: it is implicit in one's activity. The role of education cannot, because of this, lie solely in the hands of the Party, rather a new unity needs to be achieved in which the Party works to raise the good sense generated by subaltern groups. Here Gramsci advocates a new unity between a philosophy of praxis, education, and the Party. The Party becomes a functionary developing the insights of the subaltern peoples to a new level in the form of philosophy: "The active man-in-the-mass has a practical activity, but has no clear theoretical consciousness of his practical activity, which nonetheless involves understanding the world in so far as it transforms it."[26] Theoretical consciousness must be developed, in Gramsci's opinion, by organic intellectuals of the working class or the Party, who, nonetheless, function *for* the workers and not vice versa. Writing critically of the experiment with the Popular Universities in Italy, Gramsci notes that

> one could only have had cultural stability and an organic quality of thought if there had existed the same unity between the intellectuals and the simple as there should be between theory and practice. That is, if the intellectuals had been organically the intellectuals of those masses, and if they had worked out and made coherent the principles and the problems raised by the masses in their practical activity, thus constituting a cultural and social bloc. The question posed here was the one we have already referred to, namely this: is a philosophical movement properly so called when it is devoted to creating a specialised culture among restricted intellectual groups, or rather when, and only when, in the process of elaborating a form of thought superior to "common sense" and coherent on a scientific plane, it never forgets to remain in contact with the "simple" and indeed finds in this contact the source of the problems it sets out to study and resolve? Only by this contact does a philosophy become "historical," purify itself of intellectualistic elements of an individual character and become "life."[27]

Elsewhere, in more elegant prose, he writes:

> If the relationship between individuals and people-nation, between the leaders and the led, the rulers and the ruled, is provided by an organic cohesion in which the feeling-passion becomes understanding and thence knowledge (not mechanically but in a way that is alive), then and only then is the relationship one of representation. Only then can there take place an exchange of individual elements between the rulers and ruled, leaders [*dirigenti*] and led, and can the shared life be realised which alone is a social force—with the creation of the "historical bloc."[28]

Writing not long after the Bolshevik revolution, Gramsci's position on the Party retains a vanguardist element, even if the vanguard is one that must remain organically linked to the working masses. Thus, he writes of how "innovation cannot come from the mass, at least at the beginning, except through the mediation of an *elite* for whom the conception implicit in the human activity has already become to a certain degree a coherent and ever-present awareness and a precise and decisive will."[29] However, innovation does not refer here to the sparks of critical consciousness that come through the experience of changing the world or of producing everyday nature. The innovation of the elites refers to the process of transforming implicit conceptions of the world into critical theoretical outlooks. For this reason, Feenberg argues that Gramsci's elaboration offers a "striking complement to Lukács' theory in which precisely the least developed sides of Lukács' presentation are brilliantly elaborated within a generally similar theoretical framework."[30] Gramsci demonstrates the connection between real consciousness in individuals and the class consciousness that is attainable through practical activity. As Crehan points out, this position is intimately related to Marx's distinction between a class in itself and a class for itself.[31] In this sense, Gramsci's work is crucial to the project that I have embarked upon in this book. In other respects, I think we also find within Gramsci's writings a latent sense of the transformative possibilities within the socio-natural: this sense comes out most notably in his reading of the *Theses on Feuerbach*.

Gramsci's Concept of Nature

In prison, lacking access to all but the most fragmentary of Marx's writings, Gramsci relied on quoting from memory and one or two passages translated into Italian in the works of others. Alongside two or three other works by Marx and Engels, the *Theses on Feuerbach* came to assume an immense importance within the *Prison Notebooks*.[32] Peter Thomas makes the more ambitious claim that the entire notebooks can be seen as an attempt to wrestle with the *Theses*.[33] Indeed, the *Theses on Feuerbach* seems to demonstrate certain methodological principles that Gramsci then takes forward in his broader understanding of a philosophy of praxis. In dealing with these principles, he performs several fascinating moves that I think can be taken forward in a broader understanding of nature. Of particular interest here is the manner in which Gramsci seems to open Marx's claim that human nature is an ensemble of sedimented social relations to a socio-natural reading. Implicit here is a conception of nature and society as an ensemble of historically and geographically specific unities. These insights are contained within the note "What Is Man?" Gramsci begins this note by changing the question. It seems wrong to ask what man is: more important is "what can man become?" As in the *Theses on Feuerbach*, Marx counters Feuerbach's abstract conception of "Man" with a historically and geographically situated understanding of men and women operating under a specific set of social circumstances, so Gramsci develops a process-based or relational understanding. He continues: "One must conceive of man as a series of active relationships (a process) in which individuality, though perhaps the most important, is not, however, the only element to be taken into account. The humanity which is reflected in each individuality is composed of various elements: 1. the individual; 2. other men; 3. the natural world."[34] Humanity cannot, it would appear, be divorced from the natural world. The individual cannot be understood outside the specific socio-natural relations in particular places and particular times. As with Marx,

Gramsci goes on to write that relationships between people and with environments are forged actively through work and technique. Furthermore,

> one could say that each of us changes himself and modifies the complex relations of which he is the hub. In this sense the real philosopher is, and cannot be other than, the politician, the active man who modifies the environment, understanding by environment the *ensemble* of relations which each of us enters to take part in. If one's own individuality is the *ensemble* of these relations, to create one's personality means to acquire consciousness of them and to modify one's own personality means to modify the *ensemble* of these relations.[35]

Put in other terms, through producing nature, humans and their environments coevolve. Consciousness of this coevolution emerges through active involvement in the process. This is a brilliant move on Gramsci's part and one that is merely implied in Marx's own writings. It stresses the conditions of possibility for radical change that might emerge through interactions with nature. It is at the heart of the concerns within this book. Elsewhere, Gramsci refers back to this conception by broadening not the sixth of the *Theses on Feuerbach* but the third, in which Marx criticizes crude materialism for failing to recognize that "it is essential to educate the educator."[36] Thus, writing of the cultural environment, he notes that "the environment reacts back on the philosopher and imposes on him a continual process of self criticism,"[37] and, criticizing Bukharin, "If the environment is the educator, it too must in turn be educated, but the Manual [Bukharin's *The Theory of Historical Materialism: A Manual of Popular Sociology*] does not understand this revolutionary dialectic."[38] Clearly nature and society can be conceived within this framework as internally related moments in a process of mutual coevolution.

While it would no doubt be false to assume that Gramsci holds the same conception of environment as that mobilized within this book, I do think it is possible to take forward some of the more suggestive insights and to broaden them to a socio-natural reading of hegemony within Gramsci's work. Opposing what he (only partially correctly) presumes to be a separation of nature and society in the work of Lukács, Gramsci writes of how "human history should be conceived also as the history of nature (also by means of the history of science)" and, therefore, "how can the dialectic be separated from nature?"[39] The environment—understood as a differentiated unity of the social and the natural—is one specific terrain over which conceptions of the world are consolidated and contested. This, in turn, can be applied to an understanding of environmental politics and of the possibilities for broader revolutionary change. Nevertheless, in spite of these richly insightful comments, Gramsci's conception of nature, as with his understanding of the party, is far from a unitary one. Thus, on other occasions, he writes of how "the passage from necessity to freedom takes place through the society of men and not through nature."[40] Here he would seem to contradict his own comments on Lukács. Although noting that there is no fully elaborated concept of nature in the *Prison Notebooks*, Benedetto Fontana summarizes five key aspects: nature as undifferentiated matter; nature envisioned as "second nature"; nature as the irrational, as instinct, and as impulse; nature as chaos or disorder; and the potential overcoming of the domination or conquest of nature.[41] Elaborating on these aspects, he makes a compelling case that for Gramsci "man's interaction with nature mediated through labour and technology initiates the historical process through which humanity achieves consciousness of itself and its manifold relations with the world."[42] Nevertheless, Fontana asserts that this relationship is one in which humans dominate nature. Thus, in achieving consciousness of humanity's domination of nature, we gain a sense of our domination of one another, as well as the technological and liberatory possibilities for organizing our world otherwise. There is indeed ample evidence to suggest that such a conception may be present within Gramsci's work. Thus, he writes:

> One might say that the typical unitary process of reality is found here in the experimental activity of the scientist, which is the mode of dialectical mediation between man and nature, and the elementary historical cell through which man puts himself into relation with nature by means of technology, knows her and dominates her. There can be no doubt that the rise of the experimental method separates two historical worlds, two epochs, and initiates the process of dissolution of theology and metaphysics and the process of development of modern thought whose consummation is in the philosophy of praxis. Scientific experiment is the first cell of the new method of production, of the new form of active union of man and nature.[43]

Not only does Gramsci assert a domination of nature here, but he feminizes nature in a way that appears crude and ahistorical: without a sense of the ways in which domination is rooted in a historically and geographically specific way of organizing society, such a statement becomes deeply problematic. Elsewhere, he develops this understanding of the domination of nature, turning it into an understanding (not dissimilar to Lefebvre's, as we will see later) as a means through which society has liberated itself from the crises of the past. New, socially produced crises, such as the Great Depression of the 1930s emerge in the wake of this. Thus, in a note on progress and becoming,

Gramsci links such socially produced crises to a crisis of the idea of progress and to the domination of nature: "The crisis of the idea of progress is not therefore a crisis of the idea itself, but a crisis of the standard bearers of the idea who have in turn become a part of 'nature' to be dominated."[44]

However, interesting as this may be, Gramsci's writings on the domination of nature are contrary to some of the more nuanced, situated perspectives found elsewhere in the *Prison Notebooks*. Nevertheless, even within the last quotation, there is also a sense of the new possibilities that might emerge from interactions with the socio-natural world—outlined here through the process of scientific experimentation. It is with this that Fontana concludes his paper. Through coming to recognize the domination of nature in their acts, men and women may be able to forge new relationships based on socio-ecological harmony. Reversing the master–slave relation, we might seek a new ecological consciousness. This is an interesting point. But I am not convinced that it needs to rest upon any sense of the *domination* of nature. Within Gramsci's writings, there is ample evidence that consciousness of the socio-natural domain may emerge through everyday interactions with nature that have nothing to do with domination. Furthermore, given Gramsci's apparent insistence that nature and society be viewed as different moments within an unfolding totality, it would be inconsistent to argue that an integral conception of the world can emerge from anything other than this socio-natural world. Practical activity is fundamental to the making of this differentiated unity, just as it is fundamental to the generation of new ideas out of which the world may once again be transformed. Just as thought and action are internally related moments within the socio-natural whole, so human and nonhuman are mutually constitutive, coevolving parts.

While these debates may have some significant academic interest and further the development of the overall argument in this book, in Gramsci's writings theorizing is never done for its own sake. Much of the specificity of his approach lies in his ability to work through his ideas in concrete historical and geographical situations. This is the method he continually employs in the *Prison Notebooks*, and it is an integral part of what might be termed a Gramscian approach. For these reasons, in what follows, I turn to a more extensive discussion of struggles over water in Durban (and the most extensive discussion of this case within the book). Through grappling with the concrete conditions in which a politics emerges through struggles over water, we might extract worthwhile principles for the overall project of theorizing an everyday environmentalism.

Struggling for Water, Struggling for a New Civilization

Over the last half century, the informal settlement of Inanda has become increasingly integrated into the lived environment of Durban.[45] Over this period, it has transformed from a quiet rural community of dispersed farmsteads to a dense and bustling home to almost two hundred thousand people. As a liminal zone, Inanda saw some of the most violent turf wars between the Inkatha Freedom Party and the United Democratic Front in the latter years of apartheid.[46] These struggles were transposed onto generational conflicts, as the youths of the area struggled for a democratic vision of a world beyond the conservatism of village headmen and chiefs.[47] The area thereby encapsulated many of the most important changes of late apartheid.

In one of the few scholarly papers written on the settlement, David Hemson refers to the community's ability to capture this political zeitgeist: "French political life served for Marx, and in a sense for all historians and social scientists, as a barometer of all modern politics in its usage, language and innovations. Taken in proportion, and obviously to a lesser degree, Inanda shows more explicitly the latent potentialities and contradictions in political practice on a national scale by exhibiting the extremes in their development."[48] The story of water provision in Inanda is one way into making better sense of these potentialities and contradictions. For most of the last century, Inanda lacked any formal water network: its burgeoning population relied on individual landlords or shop owners to drill boreholes.[49] As the settlement increasingly came under the influence of the urban region, debates were ignited over who was responsible for the provision of water. A drought between 1978 and 1980 increased the urgency of these debates.

Lying outside the formal authority of the apartheid state and the KwaZulu legislative assembly, Inanda's administrative incoherence provided ample opportunities for apartheid governments to delay in providing water to the area, claiming not to be responsible for the area's nonexistent infrastructure. In addition, it opened up the possibility of exploiting some of the tensions between "racial" and class groups that had been exacerbated by the drought of the late 1970s. Landlords represented the key figures of authority for much of the last century. These class dynamics were, however, deeply entwined with, and inscribed by, "race": Inanda was one of the few areas in apartheid South Africa in which both Africans and Indians were able to acquire freehold land alongside one another. The preponderance of Indian landlords seeking to make a living out of "shack-farming" provided a lever from which the apartheid state was able to exploit "racial" tensions.[50]

Inanda's confusing patchwork of administrative responsibilities was partly a result of the 1936 Land Act. This act earmarked several of the local farms as Released Areas 33 and 34 for eventual acquisition by the state,[51] in preparation for sole African occupation of the area at some undisclosed future date. In the interim, the area lay in an administrative limbo, permitting both Africans and Indians to acquire freehold land. Lying in the mist belt, nearly all of Inanda is just too high to form part of the lucrative sugar lands situated to the east. As a result, the lower-value land and the high number of landowners among whom it is distributed served to ensure that, throughout the apartheid years, it retained this ambiguous, "released" status; the Verulam Magistrates Court being thus established as a temporary authority.[52]

Urbanization in Inanda was intimately bound up with the unfolding of apartheid in the rest of the country and, above all, with the forced removals of so many people from relatively stable communities in the center of cities to the outskirts of urban areas. Thus, with the destruction of Cato Manor in 1959 and the forced removal of all its African and Indian residents, new townships were constructed on the outskirts of Durban to accommodate the displaced population.[53] The construction of these separate townships for those of different "races" was the cornerstone of the National Party's policy of "separate development." Inanda's neighbor, KwaMashu, was soon to become the second largest of these townships in the Durban region, located roughly twenty kilometers to the northwest of the city center. Its creation had a profound catalyzing effect on the growth of Inanda: with the extension of bus routes to the township, it became possible for residents of the informal settlement to commute to work in the city, although pass restrictions effectively barred many of its residents from applying for such work.[54] While a basic understanding of the influence of the "structured coherence of the urban region" would demonstrate the city's ability to draw upon a surrounding labor supply, the extension of the water network to Inanda represented a still more profound connecting of the informal settlement with the processes that compose the urban region.[55] With this change, the city became a part of the home life and the bodily functions of those living in the informal settlement. A domesticated nature, embodying and expressing capitalist relations, thereby came to flow through the shacks and bodies of Inanda's residents.[56] Nevertheless, water planners had to contend with the patchwork of land titles making up the area and the lack of a clear administrative authority. At the time, Inanda was divided up between several hundred private landowners, mission lands, state lands, and land administered by the KwaZulu government. It represented a planner's nightmare. This situation was worsened by the apartheid government's plans to turn Inanda into a site for solely African occupation. This move would require the appropriation of vast swathes of land from numerous different landowners.[57]

Such a patchwork of land titles, as well as the racist priorities of the apartheid regime, was reflected in the lack of provision of basic services in the area until the early 1980s. If these were provided at all, landlords tended to be responsible. Benevolent landlords, and those with an eye for a commercial opportunity, supplied water through boreholes; otherwise residents collected water from the polluted streams that flowed through the area.[58] When these sources of water ran dry in 1979, the inadequacies of service provision were clearly emphasized. Although "silent killers" such as dysentery had affected residents for decades, a dramatic increase in typhoid cases in January and February 1980 began to draw attention to Inanda's service deficit. Fears of a contaminated labor supply began to grow, as did fears that laborers might bring typhoid with them to the center of Durban. Garth Seneque quotes an article from the *Sunday Tribune:* "Durban is sitting on a typhoid bomb. Thousands of workers could be unwitting carriers of the killer disease—and there are fears that the epidemic could spread to the city. The bomb is Inanda and the solution is water piped within easy reach of every home. But Inanda—one of Durban's major labor sources—is a squatter's haven."[59]

If the solution had been a simple one (domesticating nature through bringing potable water to the informal settlement), the responsibility for implementing such a program remained less straightforward. For the national government, the legitimacy of moves toward "separate development" appeared to be put to the test. Whichever department was made responsible for a water scheme, and whether it was a part of a government in Pretoria or a legislative assembly in Ulundi, became crucially important.

One of the key players in the drive to bring a water scheme to Inanda was the Urban Foundation. Established by wealthy industrialists soon after (and in response to) the Soweto uprising of 1976, the Urban Foundation served as a development agency, aimed at focusing capital's energies on South Africa's urban problems and transforming state policy.[60] For Bond, the organization served a vital function in temporarily displacing a crisis of overaccumulation in the South African economy through diverting funds to black housing projects.[61] For Smit, however, from the late 1980s, it became clear that transformative potentials were available within the organization and these might synergize with the emerging democracy.[62] With the Urban Foundation's formation being so closely linked to a renewed wave of opposition to apartheid in the townships, it became an important source of debate as to whether these possibilities were real or simply an attempt to enroll liberal academics in capital's hopes for an elite transition from apartheid to neoliberalism. Although there was a wide diversity of opinion within the organization, ranging from the egalitarian to the Thatcherite and outright racist, the Urban Foundation was persistently open to the criticism that it sought to make "apartheid liveable." At a time when the national government in South Africa was attempting to ameliorate some of the worst living conditions in the country as an overt strategy of attempting to quell rebellion (part of the "total strategy" in countering the "total onslaught"), the Urban Foundation appeared to be a willing (if critical) handmaiden. Whether one is cynical about the motives of the Urban Foundation or not, the introduction of a water scheme to Inanda would utterly reconfigure the settlement's relationship to both the municipality and the apartheid state.

Nevertheless, such a transformation was still effectively blocked, while landowners remained the most important power bloc in Inanda. The patchwork of land titles caused problems for planners; the coexistence of Africans and Indians also seemed to challenge the architecture of apartheid; and the local political significance of landlords as service providers provided still a further obstacle. The apartheid state was confronted with several possibilities. First, it could seek to appropriate the land. At the time, however, with the economy in recession, the state's lack of financial resources limited the chances of its being able to do this legally. Second, it could attempt to undermine the power of the landlords and drive them out of the area through extralegal means. As I will show in the next section, this strategy was evident in the manner in which tensions between Indian landlords and their African tenants were exploited by the state in the run-up to violence that engulfed the area in August 1985. Third, the apartheid government could relocate people from areas in which landlords were the dominant force to areas in which the state already had a strong presence. Essentially, this was the strategy pursued: the drought in the early 1980s provided a justification for the movement of large numbers of people from freehold land to state-owned land. The latter had been purchased from individual landowners for the purpose of integrating Released Area 33 into the KwaZulu Bantustan at some, as yet unknown, future date. To much international interest, the first "site and service" scheme was thereby established. Residents were provided with a plot of land at a reasonable rent (with the possibility in the future of being able to purchase this land), a standpipe, and the option of borrowing a tent until they had been able to construct their own house on the land. During the first phase of resettlement, three thousand families, or roughly fifteen thousand people, moved to what would become known as Inanda Newtown. Since the establishment of Inanda Newtown, three new divisions have been established—Newtown A, B, and C.

Moving residents from land in which they rented plots to an area owned by the state served several functions. First, through state-planned, speeded-up urbanization, it permitted the rapid implementation of a water scheme, avoiding potential planning delays. Second, while not involving forced removals, it served the apartheid vision of separate development for those of different "races." This relied on sole African occupancy of certain parcels of land. Third, it broke the power bloc of local landowners. Seneque refers to the changing position of the prominent Inanda landowner, Rogers Ngcobo, whose cautious support for the water project changed to opposition as he saw the loss of power that would result from resettlement.[63] Finally, the supply of water to a resettled population in parts of Released Area 33 allowed the state to rework the political ecology of the settlement. Prior to this, as Doug Hindson notes, landlords had been responsible for providing not only basic services but also "a worldview" for the local populace.[64]

This sense of a worldview being provided by local elites does not seem to ring quite true to the autonomous currents that still flow through the area. Instead, in providing services, landlords helped to shape the terrain over which worldviews emerged. In Gramscian terms, they sought to foster a worldview that frequently contradicted the implicit conception of the world expressed in people's actions. One crucial terrain over which these ideas came together is around the production and consumption of basic resources and, in particular, people's ability to access clean drinking water. Thus, if it is wrong to suggest that landlords directly shaped the consciousness of local people, it is important to recognize the ways in which relations with water—enacted through the provisioning of a household with this basic need, and in part controlled by both landlords and tiers of the apartheid state—performed a central role in condensing and articulating specific conceptions of the world.

In spite of the development of the Inanda scheme, the apartheid state's ambitions for the settlement continued to be frustrated by the patchwork of land titles in areas not owned by the state. In particular, the presence of both Indian and African residents still provided a challenge to the violent segregation that constituted apartheid. One of the expressions of this coexistence had been local peace camps—organized by Rick Turner, a Durban philosopher-activist (assassinated in 1978) and Mewa Ramgobin, a prominent member of the Natal Indian Congress and later a United Democratic Front (UDF) activist. Ramgobin is married to the granddaughter of Mahatma Gandhi and the camps, designed to promote nonracial harmony, were held at Gandhi's former home in Inanda, known as Phoenix. For some, these meetings are still remembered as key platforms for building a progressive politics in Natal.[65] Nevertheless, at the same time, as pockets of hope showed the possibilities for a society beyond apartheid, the central state was making a concerted effort to open up new fault lines. Heather Hughes charts these fractures in trying to get to grips with the violence that took hold of Inanda in August 1985.[66] Hughes's writings capture the important interplay between the apartheid state and the historical geographies that make Inanda the distinct place it is. Through doing this, she seeks an explanation for the ugly violence that took hold of Inanda. The spark was provided by uprisings in other parts of the city, provoked by the murder of UDF activist Victoria Mxenge. In Inanda, these uprisings took the form of intense clashes between an emerging, but by no means coherent, UDF and armed Inkatha militias. Many of the attacks were targeted at Indian landowners and, within a few days, the vast majority of the Indian population of Inanda was driven out of the area, fleeing to nearby Phoenix or Redlands.

As Hughes's writings attest, the roots of the violence were complex. For Sitas, the bloodshed needs to be situated in the political-economic context of the moment.[67] Much of the violence was an attack on what were perceived as symbols of wealth, following the acute crises of drought and resettlement. Also, Inkatha was able to capitalize on such violence, attempting to consolidate its power through the ensuing clashes while portraying its supporters as the peacemakers and upholders of law and order. Above all, Sitas suggests, the violence showed the limitations at that time of all organizations in the Durban area, whether of the UDF, Inkatha, or the growing trade-union movement. There seemed no way in which the embryonic postracial politics of Phoenix could cohere into a total and integral worldview that might have been capable of building the society envisioned in people's actions. Instead, a variety of morbid symptoms—racism, violence, and murder—expressed a deep instability in the emerging worldview.

Elsewhere, Fatima Meer interprets the 1985 violence as a state-sponsored effort to rid Inanda of Indian landowners.[68] With the state lacking the money to pay a reasonable price for this land, it was convenient for Indian landlords to be driven out through violent insurrection. Through firsthand reports and signed affidavits, Meer shows the manner in which the apartheid state operated as a hidden hand feeding the violence. In contrast, Hughes argues against conspiracy theories and instead alludes to tensions over the allocation of basic resources.[69] Here, again, we begin to see how everyday acts of provisioning a household with water serve as a terrain over which implicit conceptions of the world begin to emerge and articulate with the fragmented terrain of common sense. Crucial here is the drought of 1979–82 and the state's cynical response to it. After the drought, state authorities targeted Indian landlords for not having provided water supplies to their tenants, while their African counterparts were not questioned. The state attempted to disrupt any embryonic postracial stability through naming Indian landlords as authors of the degraded living conditions in the area. It was they who prevented their tenants from accessing that most basic resource. It was they who had forcefully alienated the majority from their conditions of existence. Of course there was some truth in this, but at the same time it exposed little of the roots of the crisis in the apartheid distribution of land and the refusal of the state to service the needs of the majority in favor of the few.

At the same time, further lines of tension emerged with the construction of Inanda Newtown in 1982. Xhosa-speaking residents of Inanda were effectively barred from residency in Inanda Newtown, as the apartheid state considered the development to be part of a move toward an "ethnically" distinct KwaZulu Bantustan. Thus, with potable water being supplied to Inanda Newtown, clashes occurred at standpipes in the site and service scheme between Zulu- and Xhosa-speaking residents of Piesang River. By the time Inanda ignited with violence in August 1985, the state had already, in Hughes's words, "discovered and forcibly widened every faultline in the social makeup of Inanda."[70] The fragile dominance of the landlords and the embryonic postracial worldview from Phoenix had both been shattered. New violent conflicts exploded in response. With neither landlords nor the civics, neither the Inkatha Freedom Party nor the apartheid state, able to exercise true leadership, the result was violent and racist clashes.

Even if the violence in Inanda was provoked by the state with the intention of simplifying the patchwork of land titles across the area, this strategy clearly failed. Similarly, if the relocation of residents to Inanda Newtown is considered to have been an attempt to develop greater quiescence to the apartheid regime, this also failed. Instead, opposition seemed to be stoked. Part of this can be read through the fact that the process of urbanization, as was happening in and around Newtown, often seemed to bring about a denser network of opposition to apartheid. Support for the UDF had always been greatest in urban areas, so the apartheid state walked a very narrow line in constructing such spaces with the intention of governing them. Similarly, through attempting to reconfigure residents' relationships with the provision of water, the apartheid government actually provided new tools with which the regime might be challenged. New terrains of political praxis were opened up through the creation of new practices around water provision. In part, we see these terrains being actively produced through the relations that constitute the lived environment of the settlement. Centralized provision of water in the area sought to bring apartheid subjects into a subordinate relationship with the state, the ultimate expression of which was the passive acceptance of the regime through unquestioned payment for urban services. As we will see later, the cultivation of a "culture of payment" remains a fundamental mission of the state in its postapartheid form. Nevertheless, by providing centralized services, a new weapon was given to local residents around which opposition to apartheid could begin to cohere. A payment boycott could thereby be organized, the intention of which was to challenge the apartheid state's stronghold. For Gillian Hart and others, a key moment here was the establishment of the Black Local Authorities (BLAs) by the apartheid state, which effectively galvanized opposition to apartheid and became a central target around which civic groups could form.[71] Crucial to the transformation of the new water network, from a tool for the consolidation of apartheid into a weapon for the challenging of apartheid, was the urbanization of opposition that was expressed in the emergence of new civic movements. Although the emergence of the Civics is fundamentally related to opposition to the formation of a tricameral parliament in South Africa,[72] this understanding must also be supplemented with a reading of how struggles over land and water provided much greater coherency—and a far more direct, everyday concern—for these new movements. Again, the terrain over which the Civics sought to challenge apartheid was a political one, but it was also one intimately related to the production of apartheid within the lived realities of the majority of the population. Necessarily, these lived realities need to be situated within the political ecology and historical geography of the area. If, as both Marx and Gramsci were well aware, relations to nature are one moment in a dialectical totality, this should lead to a questioning not only of how that totality comprises these distinctive moments but how these moments come to embody the changing moments elsewhere.[73] Apartheid was expressed in specific relations with nature, just as these socio-natural relationships provided further weight to the overall apartheid project. Challenging this meant a battle on many different fronts, one of which was people's ability to access clean drinking water.

The Emergence of an Oppositional Force

In Natal, the "comrades phenomenon" made this opposition to apartheid distinctive. Ari Sitas refers to the comrades as "a large scale social movement with its peculiar Natal overtones,"[74] juxtaposing this with the commonplace assumption that the youths represented a suicidal expression of a loss of hope in the potential for change. In Inanda, the comrades seemed to focus the energies of a new generation of progressive youths. While many of their aims were short-term, the movement also seemed to contain the seeds for a potentially lasting transformation of the socio-political landscape.[75] Not only did the comrades become a crucial counter-hegemonic force within Inanda—thereby cultivating a new set of moral and cultural norms[76]—but with the gradual collapse of authority in the area, they served as agents in reworking relationships with the lived environment. They were insurgent architects of a new relationship with nature and a new political ecology through the various democratic planning forums established and the distribution of land to needy parties. With this, new implicit conceptions of the world could emerge from within the common sense of the area. Potentially these incoherent insights could be brought together in a new fusion of theory and practice through the transformation of the "rational kernel" within. Temporarily, the comrades replaced both traditional authorities and apartheid structures of governance, providing a grassroots, participatory channel for reshaping the environment of the settlement.

In the Amaoti area of Inanda, one of the sites worst affected by the 1979–82 drought and one having experienced a major disruption of hegemonic formations, the comrades played a vital role in distributing land and housing. While there was clearly the potential to abuse this position in order to cultivate patronage networks, Preben Kaarsholm quotes interviews showing that many comrades were so restrained as to actually leave themselves with no home at all.[77] This disciplined altruism was reflected in the formation of the Inanda Marshals by a few key comrades such as Thulani Ncwane and Sabatha Ngceshu.[78] This grassroots movement of youths was committed to the ending of apartheid and the establishment of peace: David Hemson describes it as having combined spontaneous self-organization and a militia formation.[79]

The character of the Inanda Marshals was quite profoundly misunderstood by the returning exiles from the ANC, who appeared to subscribe to the dominant stereotype that the comrades were nihilists. Hemson refers to Nelson Mandela's speech in Durban in 1990 as a defining moment through which the Marshals realized their distance from the Congress leadership. This distance was then reasserted when, in March 1990, a meeting was convened in Inanda with the proposition being put forward by the emerging local leadership of the ANC to dissolve the Marshals and replace them with the ANC Youth League (ANCYL), a more easily contained "civic movement."[80] Thulani

Ncwane describes the meeting: "If a democratic resolution was passed to dissolve the committees and join the ANC Youth League then we would do this. But instead, this guy came along with a shirt, tie and gold watch and brief case. In the past we had only seen Dubes dressed like this. And then he said, I am from the ANC Youth League convening committee and we have come to organise your local branch."[81]

The imposition of the ANCYL was part of a broader attempt by the Congress leadership to formalize disparate civil society movements across the country within one hierarchically organized structure. The autonomy of the Inanda Marshals, as architects of a reconfigured political landscape within Inanda, posed a potentially unwanted source of opposition. For the Marshals, this conflict was both unnecessary and debilitating. Thulani, for example, saw no division between the work he was carrying out within the ANC and the work he was carrying out with the Marshals.[82] In spite of this, in 1992, he and another Marshal were suspended from the ANC for alleged Trotskyist tendencies. This was a clear attempt by the Congress movement to target dissenters in an effort to better facilitate the cultivation of nationalist authority. By mid-1992, the structures of the Marshals were in disarray.[83]

Activists such as Ncwane now turned their work into more formal movements, groups that were more readily recognized by the returning ANC leadership but still had a relative autonomy. With the first democratic elections in 1994 and the wider political transition taking place in the country, civic movements were formalized further. At the time, however, such groups still remained relatively autonomous within Inanda. Thus, the Inanda Civic Association (ICA) flourished and, in 1994, the Inanda Development Forum (IDF) was launched by a group of landowners, local politicians, and civic activists. Unique in its structures and aims at the time, the IDF attempted to bring together disparate groups across Inanda that had formerly been in conflict. In particular, the Inanda Landowners Association and the Inanda Civic Association, two organizations normally opposed to each other, were brought into the conversation. As a result of this new dialogue, various innovative structure plans for Inanda could be devised through the IDF, which also sought to train youths from both within and outside the comrades movement as community development workers. One community development worker was appointed from each of the thirty-two communities composing Inanda and it finally seemed like the energies of previous generations could be directed at reconfiguring the lived environment of the area. Through this reconfiguration, a new worldview might be fostered, based on the embryonic and implicit conceptions developed in the dying days of the party and now articulated with participatory modes of governance and efforts to deepen direct democracy. With no formal government in Inanda, different local and national administrations respected the IDF as a semiautonomous local authority.

By 1996, with the gradual development of a democratic system of local government in South Africa, this situation was clearly no longer acceptable: the Inanda Development Forum appeared to provide an unwanted and unnecessary power base, blocking moves toward wall-to-wall local government. The embryonic worldview it was nurturing ran counter to the now-centralizing tendencies of the ANC. With the appointment of Thulani Ncwane as chairperson, the likelihood of dissenting views was once again brought to the fore. The ability to maintain Congress threads within the IDF seemed far less feasible and, again, the ANC turned to establishing more effective channels in order to consolidate its own moral and cultural leadership. In particular, the party called for the establishment of ward committees to work in close collaboration with local councillors and ostensibly provide a forum for open debate about the needs of the local area. Municipal funding to the IDF was removed and, by 1997, this organization was also in crisis. In the first local government elections of 2000, several of the community workers who had been trained and nurtured within the IDF became ANC local councillors.

As this brief review of the formation and rebirth of civic moments within Inanda has shown, repeatedly the settlement has seen the appearance of vibrant autonomous movements; and repeatedly they seem to have been either disbanded or attempts made to quash their autonomy. The result seems to be a considerable level of political disillusionment and mistrust for those in authority. Election turnouts, although still high in comparison to the UK, are beginning to diminish. While the ANC would appear to have consolidated power within these formal institutions of government (Mshayazafe remains the only "Inkatha area" in Inanda), this control would appear to have been gained at the expense of channels through which dissent might be expressed. The Inanda Civic Association is now a hollow shell, the Inanda Development Forum has become an organization only in name, and the various ward committees established for devolving democracy to the grassroots appear to serve merely as forums for cultivating the next generation of councillors.

It is not only for dissenters that this situation poses problems. Indeed, if organs of civil society have been lost, the potential for cultivating consent is also reduced. Kaarsholm suggests that conflict is now centered on various "moral panics" because people are unable to find outlets for their anger in a political arena almost entirely monopolized by one party.[84] Crucially, however, although seemingly omnipotent, the ANC has actually lost many of the "networks of tunnels and earthworks" that Gramsci recognized as being so crucial for defending particular sociopolitical regimes.[85] It has virtually lost its Youth League in the area, it no longer has strong links with the civic association, and, with growing disillusionment at the pace of reform within postapartheid South Africa, the likelihood increases that the ANC's ability to *lead* in Inanda will come under serious threat. This situation poses serious predicaments for those concerned by South Africa's elite transition. On the one hand, this partial hegemony would appear to be undergoing a precarious transformation. On the other, sources of democratic opposition seem to be thin on the ground: those movements that do exist are frequently disinterested in a direct challenge to class power. However, if there are few obvious signs of overt forms of class struggle, I would suggest that there are tentative signs of a war of position emerging within Inanda. However, this takes a different form from that classically envisioned within Gramscian scholarship: it appears to be conducted as much through the lived environment and the technologies through which relations with it are mediated. These have been rediscovered as moments in which latent struggles over power and dominance might be fought. Through seeking to forge new sociotechnical relations with the lived environment, mental conceptions, everyday life, and broader social relations are in turn transformed. Nature, and the relation that those living in Inanda have with nature, is one moment in this coevolving socio-natural totality.

From Apartheid to Postapartheid Struggles for Justice

Most obviously, new partnerships have formed bringing together the state, a very broadly defined "civil society," and the private sector. Through these partnerships, as in the apartheid years, relationships of power are produced through the everyday acts that connect people with the urban water network. This network, and the stability being afforded to a worldview through it, now encompasses the entire settlement. With the recent commercialization of bulk-water supplies to the municipality, the infrastructure through which that water flows has come to acquire a life of its own. More clearly now, it embodies and expresses the apparently alien dominance of capital over people.[86] With these reshaped relationships, it has become both easier and more important to ensure a modicum of consent to the apparent dictatorship of the water meter. One vehicle for fostering such consent has been the Business Partners for Development "learning partnership," piloted in Inanda as part of an effort to provide better services to the area and, importantly, to tackle the high incidence of nonpayment for water services. It is to this, and to some of the hopeful signs that this war of position will be lost by capital, that I now turn. These emerging possibilities can only be understood, however, through their articulation with the preceding historical geography of the settlement.

The postapartheid government has seen one of its primary challenges in the delivery of services as being to transform an alleged "culture of nonpayment" within poor communities. In "tackling" this mirage, municipal bureaucrats have conducted an all-out cultural bombardment of poor communities, seeking to cultivate what they perceive to be responsible, fee-paying individuals. While such discussions nearly always overlook the question of people's inability to pay for these services, it has reopened the water network to the contestation of ideas (expressed and embodied in everyday acts) already discussed. The Business Partners for Development (BPD) (or "Building Partnerships for Development" as it has since been renamed) initiative must be seen as one aspect in this struggle to transform common sense. BPD claims to be "a worldwide network of partners involving government, donors, business and civil society."[87] Created by the World Bank in 1998, it has helped to coordinate trisector partnerships between governments, "civil society," and the private sector throughout the world.

The water and sanitation cluster of the group has pioneered twelve local-level projects, one of which was based in KwaZulu Natal—Inanda being one of the two focus areas in the province. Thus Inanda became an important testing ground for an international drive toward such trisector partnerships. Here a project evolved between eThekwini Water Services, Mvula Trust (an NGO specializing in the provision of water to rural communities), and Vivendi Water (now Veolia and one of the two largest multinational water companies). The aims and results of the partnership remain somewhat ambiguous: much of the time the project appeared more like a period of courtship between the main partners, prior to what many hoped would be a private concession contract. Tangible results are minimal—amounting to little more than a slight change to the municipality's ground-tank system and steps being made toward the piloting of a new sewerage system for informal settlements. The self-acknowledged principal outcome, however, seems to be the establishment of a "learning partnership."[88] In many ways, the effects of the learning partnership have been more subversive than the presumed effects of an outright divestiture; it is here that we see what might be understood as the attempt to provide new moral and intellectual leadership through both the conventional organs of civil society and through a more broadly defined "produced environment."

For eThekwini Water Services, one of the principal lessons the municipal entity claims to have learned from Vivendi Water is how to deal with "customers."[89] The language used in both interviews and in the BPD documentation is quite explicit about residents of the municipality being customers, and a large part of the BPD seems to have been an attempt to rescript citizens as atomized, fee-paying, consumers of services. Prior to this, eThekwini Water Services clearly felt compromised in its dealings with "customers." Interviewees from both Umgeni Water and the municipal water provider commented on the political astuteness of the South African consumer,[90] and this seems to have served as both a powerful source of opposition to water privatization and a source of discontent over attempts to introduce cost-recovery initiatives.[91] Locally, the agitation of the Concerned Citizens Forum,[92] the higher-than-average profile given to the Christina Manquele court case that challenged eThekwini Municipality's reading of the right to water,[93] and the apparent willingness of the council to side with the "poor black majority,"[94] have pointed to major failings in eTWS' approach to billing and constraints to its ability to impose water meters and disconnect supplies. The killing of two disconnection bailiffs is one of the more overt signs of the limits to the municipality's more conflictual approach to imposing charges on poor residents (involving an incredible four- to five thousand household disconnections per week at the time of research). Now, the head of eThekwini Water Services seems to argue for a slightly more conciliatory approach to ensuring that bills are paid. A two-pronged strategy, he argues, is necessary. One aspect of this strategy is to combat the agitation of the social movements organizing around water issues and win back the support of the council. The key to this, Neil Macleod argues, is to work on the attitude of the Exco, the cabinet committee of the municipality.

The second aspect is to overturn the "culture of nonpayment" that Macleod assumes to be the haunting specter at the heart of all problems experienced by eTWS. As well as the tighter quantification of supplies referred to previously, the strategy to be developed to counter this lies in the more "decentralised" approach "learnt" from Vivendi:

> We've learnt a lot from Vivendi. . . . Vivendi have a particularly refreshing approach to customer management. At the moment the Metro is communicating a lot with its customers but this communication is not working. It's important to establish a person in an area responsible for community liaison. This will lead to more empowerment. We're now thinking of employing somebody working in a [ship] container. My hope would be for more private sector involvement of this kind in future.[95]

Perhaps not surprisingly, a representative from Vivendi feels the same, arguing that the company's most important gift lay in fostering this decentralized approach to customer management. On the surface, this seems relatively unproblematic. An office in the community providing people with a space in which to seek information, lodge complaints, or query bills is surely vital for the provision of a quality service. However, beneath the surface, it becomes clearer that the prime motivation for this decentralized approach to customer management is not to answer technical queries but rather to ensure the regular payment of bills: "It is too easy not to pay for bills. . . . A form of "punishment," if you might call it that, needs to be quicker. And education needs to take place around this issue."[96] Interestingly, LeMaux emphasizes the importance of a combination of both the coercive and the consensual. Whereas previously the municipality had been heavy on the coercive approach to ensuring regular payment of bills, provoking both protest and anger, lessons from the private sector on the need to foster consent became available through the "learning partnership." This has intriguing echoes of the partnership with capital represented in the Inanda scheme in the early 1980s. With the channels through which the ANC had previously attempted to manipulate consent no longer being available (the civics and the ANCYL were, as I have suggested, in an enfeebled situation), the local state has worked with the private sector in pioneering new ways of reconfiguring residents' relations with the lived environment. Inanda's lived environment has thereby become one of the key terrains over which the moral and cultural legitimacy of the ANC's neoliberal turn is both consolidated and contested in the area. I do not intend this to be understood in a functionalist manner at all. Neither local nor central elites saw their role as being to produce neoliberal citizens. Nevertheless, as I have sought to demonstrate, both recognized the importance of enhancing a worldview based on individual entitlements built on the backbone of the atomized paying consumer of services. The very functioning of a water service in Durban increasingly came to depend on reconfiguring relationships to this service and, in the process, transforming understandings of it. In doing this, relations to water came to be reconfigured and served to generate a changed "culture." Again, all these moments need to be understood dialectically. If the relation with nature has been significantly downplayed within secondary writings on Gramsci, as I sought to show in the opening sections to this chapter, this is not because they are unimportant. Indeed, active relations with nature are fundamental to the formation of embryonic and implicit conceptions of the world within Gramsci's writings. Nevertheless, the battle over which group is able to fuse theory and practice in a way that produces a "coherent" worldview is also fundamentally related to nature and technology. Each of these moments embodies an internal relationship with the others. The strength of a philosophy of praxis lies in its dialectical relationship to these historically and geographically situated practices.

Apart from the attempt to shift to a more decentralized approach, the BPD project has initiated a flurry of research into the reasons for the persistence of a culture of nonpayment and initiated various educational programs prior to the installation of new water connections, urging residents to keep up regular payments and arguing the case for the necessity of charging. BPD documents state that "within the townships there is a strong culture of nonpayment and entitlement, stemming from the apartheid-era resistance movement. This low willingness-to-pay is proving difficult for the pilot to overcome and impairs cost recovery."[97] In most cases, nonpayment is, however, not a matter of choice but a matter of necessity. In spite of this, educational visits were developed in which the principal purpose was to convey to people "the advantages of paying" and the "disadvantages of an illegal connection." Mvula Trust was the partner in the BPD project responsible for coordinating educational visits. Its strategy was one of selecting civic activists to bring this message to the communities. In doing so, they gained privileged access to the community. As one of the participants, a key civic organizer in the Amatikwe area of Inanda, commented: "As we knew the advantages and what people would say and think, we were told to send the message that this [illegal connections] was not helping plumbers who would be employed in the new system, and that there were problems with burst pipes and pressure dropping"[98] The old civic structures were thus partially rejuvenated in an effort to rework the political ecology of Inanda. This micro-level passive revolution employed former activists in the hope that they could rediscover their old tunnels, excavated through periods of struggle against those in authority. This time, however, they were to be used to consolidate new choreographies of power represented in the responsible fee-paying consumer of services. Both recognizable civil society institutions and the lived environment of the settlement were brought together in the hope of guaranteeing the community's acceptance of the ANC's rightward shift.

In 2003, a second phase of the trisector partnership was launched. This time it was no longer under the umbrella of BPD and instead involved liaising directly with "civil society groups." It was given the isiZulu name Sisondela Kuwe, meaning "getting closer." Any vestiges of the partnership being about improved service delivery were abandoned at this stage, as the sole focus of the new partnership was customer management and overturning the "culture of nonpayment" through the recruitment of local activists and a vigorous education program at the micro level.

The attempt to consolidate a culture of the responsible, bill-paying consumer of services cannot of course be seen in isolation from other shifts taking place within South Africa. Principal among these must be the Masakhane campaign. This campaign stretches as far back as 1995, first appearing in the local election manifesto of the ANC before being reiterated in its manifesto for 2000. Masakhane means "build together": its aim is "to inculcate the culture of payment" and to reverse what is presumed to be a continuing economic boycott of payment for services.[99] As with Sisondela Kuwe, Masakhane is given an isiZulu name to ensure that its entry into communities is given a level of indigenization. Given the absence of "a culture of nonpayment" in all the areas in which research was conducted, neither Masakhane nor Sisondela Kuwe can, however, really be understood as challenges to a genuine culture of nonpayment. Rather, they seem to be efforts to cultivate a moral obligation to pay for what is actually unaffordable. In this regard, their aim is to develop consent within broader political economic shifts. Rather like Gramsci's interest in the Taylorist techniques of Fordism creating a new economic being, the pilot trisector partnerships in Inanda would seem to be micromanaging service delivery in an effort to bring into being a new neoliberal consumer. They seek to build from what is perceived to be commonsense understandings while also failing to grasp the conjunctural specificities. Interestingly, for Gramsci, writing over seventy years earlier, and in a classically dialectical move, rather than finding the situation of the Fordist human to be hopeless, he detects new possibilities.[100] The routinization of tasks does not mummify us as workers; it frees us up to develop new ways of thinking about progressive change.

Situating the Philosophy of Praxis within a Political Ecology

As we see through the changing historical geography of Inanda, hegemony can be rethought as a particular mix of consent and coercion that is achieved in part through historically and geographically specific relationships with nature. Just as with the philosophy of praxis, I would argue that the problematic of hegemony can be rethought to make these socio-natural relationships more explicit. While it would be simplistic to state that consent and coercion work *in* specific environments, it is far more interesting to understand how they come to be combined *through* specific environments. If Gramsci's absolute immanence is indeed concerned with what he terms the "absolute earthliness of thought," this necessarily means engaging with such politicized environments. These environments themselves change historically: in the case of Inanda, nature is gradually urbanized and, in the process, residents' relationships to water, the metropole, and the central state have been reconfigured. The environment of the colonial state gave way to a landscape scarred by deepening forms of segregation and racism. This racism was expressed and lived in the very act of collecting water from a standpipe, in the act of drilling a borehole, and in the act of challenging a councillor to provide an improved supply. Rather than necessarily always provoking anger and hostility, the granting of concessions was often seen to provide a modicum of consent to apartheid's authority. The ability to blame the reality of resource scarcity on Indian landlords permitted the fragile stability of the settlement to be fractured by the apartheid authorities as they sought to disrupt the interracialism of the area and bolster "separate development." In the present moment, the municipality has been relentless in its desire to produce a culture of payment within the community. Individuals' relationships to water have been redefined in strictly volumetric terms, where every last drop is metered and tightly commoditized. In the process, this has helped to stabilize a view of the economic-maximizing individual agent of the postapartheid period. Citizens have become consumers through interactions with the flows of urbanized nature in the community: at the same time, consent to the postapartheid project has, in part, been secured. This is not to deny the deeply coercive aspects of hegemony—families being regularly disconnected, bailiffs clashing with householders—but it is to see both moments of hegemony (consent and coercion) working within a socio-natural whole.

Of course we know that Gramsci must have viewed hegemony as, at least in part, a socio-natural problematic because of his refusal to separate nature and society shown in his nuanced dialectical conception. Nevertheless, by making the socio-natural workings more explicit, I think we are provided with a stronger material understanding of hegemony than is often permitted. This should not be understood in a one-sided determinant fashion. The delusions of environmental determinists have no place within a reworked understanding of a philosophy of praxis, nor do they have a place in reinterpreting the problematic of hegemony. Neither is there a place for the notion that consent can somehow be automatically guaranteed in a worked-over nature, as if the sole route to securing power lies in some Stalinist project of reengineering the environment. Again, nature is but one moment in the problematic of hegemony. It provides a further mediation, almost always downplayed, through which consent and coercion come to be combined. It provides a material understanding to the network of earthworks and tunnels that Gramsci describes in his evocative metaphor for civil society in the West.

But if this implies a more materialist understanding of hegemony—always present in Gramsci's *Prison Writings*, even if not always emphasized sufficiently—this is not to imply that domination and subordination are the only possible outcomes. As the example of Inanda shows, a weapon is provided in the urbanization of nature through which the apparent immutability of power relations might be challenged. On the one hand, this was evidenced through the ability of residents to target the financial stability of the apartheid regime. On the other hand, new embryonic conceptions of the world emerge within the lived environments that are produced. Throughout this book I have sought to demonstrate that the act of provisioning a household with water provides conditions of possibility for situated understandings of the processes and relationships through which environments are constituted. What happens to these situated understandings and flashes of radical thought is one of Gramsci's primary concerns. As in the violence in Inanda in 1985, unless it is possible for a progressive integral conception of the world to be vocalized in a coherent manner, such embryonic worldviews are gravely endangered. Here the question of organization, clearly so fundamental to Gramsci's overall perspective, needs to be raised. Although throughout Inanda there is ample evidence of rich, nuanced, and radical critiques of the urbanization of injustice within the settlement, these are frequently stifled by the inability to become an explicit critique. For Thulani Ncwane, this radicalism necessitated a return to the South African Communist Party. Just as the PCI seemed to provide such a vehicle for Gramsci, this was a vehicle through which implicit conceptions of the world might become coherent, explicit worldviews capable of turning grassroots theorizing in Inanda into a material world-changing political force through the fusion of theory and practice. Nevertheless, this gamble runs the risk of simply subsuming much of this radicalism within the propagandist and vanguardist stance of the Party as a whole. Elsewhere within the city, the emergence of Abahlali baseMjondolo has become one sign of the possibilities for organizing outside the tripartite alliance. While not without its own problems, Abahlali has played a remarkable role in building a radical, educative shack-dwellers' movement capable of providing coherency to the flashes of anger from within informal settlements. The xenophobic violence of 2008–9 demonstrates the necessity of such a movement more than ever. At the same time, Abahlali has become a target for those seeking to crush any sources of opposition to the ANC.

Nevertheless, as Thulani's actions showed, the ability to organize freely outside the tripartite alliance is often seriously hindered and its efficacy questionable. The ANC is fundamental to the common sense of the vast majority within South Africa. It assumes a near folkloric importance and, not surprisingly, has been able to bolster many people's understandings of the potential for fundamental change in the world. To seek to build good sense outside any organic relationship to this common sense would be to make the same error as Bukharin. His purist conception of historical materialism loses this organic relationship and seeks to build a communist moment on the back of rigid laws rather than the messiness of the existing world. Gramsci urges us to begin elsewhere, on the terrain of common sense. Perhaps, too, in the messy reality of lived and produced environments.

Conclusions

Much more than in previous chapters, I have sought to develop an argument here by grappling with the concrete realities of one particular lived environment. I have taken it as assumed that the quotidian act of producing nature opens up conditions through which embryonic conceptions of the world might emerge. This much we have established already. Instead, I have sought to question what happens to these worldviews when they articulate with conceptions that already exist and with the power relations that they potentially challenge. Underlying this endeavor is an attempt to think through how a latent volcanic anger might be turned into a coherent and slow-burning rage capable of achieving a lasting transformation of the real world. More than most theorists, Antonio Gramsci dwelt on these issues. Indeed, his life experience was, as with so many other people, profoundly shaped by the reversal of a revolutionary moment and the failure to build on the radical insights that came from within the Factory Council movement. For Gramsci, organization was clearly fundamental. In order to transform the kernel of radicalism within common sense required verbalizing and making explicit an embryonic perspective. Doing this required a detailed knowledge of the specific conjunctures through which different worldviews articulate with one another.

Although, for authors such as Foster, Gramsci is bundled up with a host of other marxists who are said to have rejected the attempt to apply the dialectical method to nature and science,[101] he was in fact acutely aware of the manner in which nature is one moment in a continually coevolving totality. We cannot expect an ecological politics within the *Prison Notebooks* (as Foster seems to): instead, it behooves us to ask whether we can meet the challenges Gramsci lays out for us with any ecological politics that we might develop today.

Through the example of Inanda, I have demonstrated how consent and coercion are always embodied and expressed as part of a socio-natural assemblage. In turn, constructing a progressive politics based on equality, respect, and dignity involves considering how embryonic conceptions of the world emerge from and articulate with already existing worldviews. Through the act of collecting, paying for, and drinking water, each of us engages in an activity that brings us into a relationship with a range of different actors and a range of different processes. When that supply of water stops, when the cost of it becomes an increasing burden or makes us sick, the politics that emerges can either remain as broken and jagged shards or be fused into a "coherent" worldview and thereby be turned into a weapon that strikes to the heart of our topsy-turvy world. Herein lies the basis for a genuinely transformative politics. How we might then turn this into the creative energy to rebuild anew is the basis for the next chapter.

Chapter 5   Cultural Praxis as the Production of Nature
Lefebvrean Natures

> Let everyday life become a work of art! Let every technical means be employed for the transformation of everyday life! From an intellectual point of view the word "creation" will no longer be restricted to works of art but will signify a self-conscious activity, self-conceiving, reproducing its own terms, adapting these terms and its own reality.
> —Henri Lefebvre, *Everyday Life in the Modern World*

> Critique of everyday life encompasses a critique of art by the everyday and a critique of the everyday by art. It encompasses a critique of the political realms by everyday social practice and vice versa.
> —Henri Lefebvre, *Everyday Life in the Modern World*

ON A COLD FEBRUARY DAY IN NORTH LONDON, City Mine(d) has set up camp in a dull and dreary cul-de-sac, sandwiched between a crown-green bowling club and a school playing field. The group's newly painted caravan, on a visit from Barcelona and recently nicknamed Gua-Gua (apparently Catalan dogs bark "Gua-Gua" and not "Woof-Woof"), is a marked contrast to the surrounding doll's-house estates in which, Lefebvre may well argue, the everyday reigns "in the chemically pure state."[1] Bemused residents have come to see what has been happening to the ping-pong balls they have spent the last week firing into City Mine(d)'s hydraulic network of plastic tubing. Through a tinny public-address system, a brusque mayor of Brent seeks to provide some clarity: he will take the balls to the head of the council, and the council will take action. Was this the clarification needed? Or was something else at work?

For a week or so, the metropolitan nature of this quotidian space had been opened up to residents in different ways. Connections were forged that previously existed invisibly: relationships between the crown-green bowlers, the children of the local school, Network Rail, and the postage-stamp gardens were brought to the fore. Less than a "moment" and more of an extended process of forging these connections, City Mine(d)'s Ping Pong Project exemplifies some of the ways in which cultural praxis can serve to rework the socio-natural ties that make city life. It opens up a mode of working that I will dwell on in this chapter. To do this, I turn to the richest and most comprehensive theorist of the everyday, Henri Lefebvre. Lefebvre's marxism has clear affinities with the position developed throughout this book. Extraordinarily prolific, Lefebvre published seventy books within his own lifetime, the majority of which are still not available in English translation. His is a marxism pregnant with possibilities, always willing to hold the present moment to account through a demand for life to be better—more joyful, more fun, more beautiful. Still writing in the 1980s, Lefebvre never abandoned this forward-looking vision: he never gave up on the possibilities for remaking the world.

Nevertheless, in spite of some suggestive comments that run throughout his work, Lefebvre was never quite able to see nature as an *ally* in the struggle for this better world. Where a concept of nature does emerge, he seems to fall back on a romantic understanding of an undisturbed nature that is disrupted by the encroachment of capitalism and modernity. Although no doubt comforting, this is also politically problematic, foreclosing the possibility of extending either Lefebvre's model of cultural praxis or his analysis of the production of space to a discussion of our sensuous entanglements with human and nonhuman others. Nature is a discrete sphere from which the splintering and rationalizing tendencies of the social should no doubt retreat. In what follows, I seek to challenge Lefebvre's conception of nature while extending his understanding of cultural praxis. The project I embarked upon in previous chapters—to make a case for metropolitan nature as a differentiated unity woven through sensuous praxis—has clear affinities with Lefebvre's revolutionary project "to extend the boundaries of praxis into aesthetic experience and aesthetic experience into praxis."[2] If, however, Lefebvre seeks to extend praxis to the environmental qualities of the city, he does so with an impoverished understanding of what constitutes that environment. Again, this is the signal move that Neil Smith performed and that was taken forward in recent writings on urban technonatures and the cyborg city.[3] It is also perhaps the key distinction between Smith's understanding of the production of space and Lefebvre's own. The city is not simply a social construction ripped from its natural hinterland and transformed into a terrain over which worldviews compete and are consolidated. Instead, it is a fundamentally socio-natural entity. Recognizing this fact not only counters some of the antinomian aspects of Lefebvre's conception of nature, but it opens the possibility for a radical environmental praxis embodied within urban interventions. This, I argue, provides a richer conception of the moment of revolutionary change.

Lefebvre, Metaphilosophy, and Praxis

Ever since Lefebvre's death in 1991, interest in his writings in the English-speaking world has grown exponentially. Initially, much of this work came from within the spatial disciplines prompted in part by Ed Soja's claim that Lefebvre was a forerunner of postmodernism, and also by the English translation of *The Production of Space*, which, with an afterword by David Harvey, radical geographers quickly picked up on.[4] Although geographers' engagements with Lefebvre clearly extend further than this momentous event, the translation of *The Production of Space* marked a dramatic shift in Lefebvre's reception.[5] Reflecting on this period, Kipfer et al. suggest that the 1990s reception of Lefebvre's writings within the English-speaking world can be divided into urban political-economic interpretations and postmodern ones.[6] Neither, they argue, does justice to the complexity of Lefebvre's thought and, as Andrew Shmuely claims, the perception of a political-economic, and a postmodern Lefebvre merely served to reproduce a division of the left into two artificial camps: with the one exhibiting a continuing commitment to the class struggle and the other celebrating difference and plurality.[7] In contrast, Kipfer et al. advocate a third constellation of Lefebvre studies, in which the rich diversity of his work is recognized. This would link "urban spatial debates more persistently and substantively with an open-minded appropriation of his metaphilosophical epistemology shaped by continental philosophy and Western Marxism."[8] Above all, it would demonstrate the relevance of Lefebvre's marxist metaphilosophy for confronting the concerns of cultural theorists, especially around questions of difference. Lacking in this third constellation, nevertheless, is a thorough engagement with Lefebvre's conception of nature.

In some ways, this third wave of Lefebvre scholarship echoes the project that Elden embarks upon in his account of Lefebvre's work. Elden seeks to show how Lefebvre can be understood only "in the context of his Marxism and philosophy more generally."[9] For Elden, more controversially, this implies a full account of his engagement with Heidegger.[10] Lefebvre's influence now once again extends well beyond the spatial disciplines and, interestingly, on the back of this, would appear to have experienced a partial resurgence in France.[11] Important for this book, Lefebvre's work has also been increasingly recognized for its importance in urban interventions and critical spatial practice: he is now recognized as offering rich and variegated insights on the moment of radical critique and his utopian sense of the possibilities for differential space within the extended field.[12] Taking this work further, John Roberts has demonstrated the fundamental importance of Lefebvre's critique of everyday life for extending and transforming already existing currents within marxism.[13]

The Production of Space

Lefebvre is still best known within Anglo-American worlds for his pathbreaking monograph, *The Production of Space*.[14] This utterly beguiling work makes a passionate claim for space as an active moment within the totality of the social. For geographers, coming alongside a so-called spatial turn in social theory more generally, this came to represent the kind of heavyweight philosophical support they needed in order to demonstrate that space matters. It is for this reason that Soja was so quick to enroll Lefebvre at the vanguard of a coterie of theorists who took on the historicism of orthodox marxism. Without doubt, this is a crucial legacy of *The Production of Space*: however, far from relegating time in favor of space, it would be more accurate to claim that Lefebvre develops a historical-geographical materialism in which time and space are kept very much alongside each other. Here Lefebvre's approach is not dissimilar to Gramsci's seen in chapter 4, or that of more recent historical geographical materialists such as Harvey. Nonetheless, if we accept that Lefebvre develops a proto-historical-geographical materialism, it remains important, especially in the context of developing an immanent critique of the nature of everyday life, to note the crucial ways in which his understanding of the production of space differs from both Harvey and Smith.[15] All too often these distinctions are collapsed and a variety of ways for approaching the production of space are rolled into one. Later, as we will see, keeping different aspects of these authors' understandings in tension permits a radical rereading of the conditions of possibility within the contemporary city.

For Smith, the production of space cannot be understood without first exploring the production of nature. Concrete acts, out of which nature is produced as a differentiated unity, are what set in motion the necessarily uneven development of space under capitalism. Smith's criticism of Lefebvre, for failing to develop an understanding of the production of nature, scale, and uneven development, as Kipfer et al. argue, is factually correct but in some ways is unfair.[16] This, after all, is the signal contribution that Smith himself makes. Within Harvey's work we find perhaps the richest political economic understanding of the production of space. Harvey's magnum opus, *Limits to Capital*, begins, therefore, by revisiting three volumes of *Capital* as a springboard for theorizing space. The result is Harvey's distinctive theorization of the production of space through spatial, temporal, and spatiotemporal fixes to crises of overaccumulation: space, in short, is an active moment in a crisis prone system of accumulation. This produced space both facilitates and provides obstacles to the ability of further crises to be displaced.

For Lefebvre, nevertheless, greater emphasis seems to be placed on an understanding of space through Marx's *critique* of political economy. The well-known conceptual triad through which Lefebvre seeks to understand the production of space foregrounds affective and mental conceptions as much as those spatial practices rooted in the political economic constitution of contemporary society. Too often this is lacking in Harvey's own work, as his own auto-critiques are quick to point out.[17] If Lefebvre asks the bold rhetorical question—"Could it be that capitalism survives through producing space?"— he actually seems far less interested in the kind of spatiotemporal fixes that Harvey describes and seems far more interested in the way in which consent is consolidated through space. Indeed, as Kipfer demonstrates, one of the signal contributions of *The Production of Space* is to urbanize hegemony.[18]

Lefebvre's Marxism

If Lefebvre's understanding of the production of space is a distinctive one, this stems directly from his open and integral engagement with marxism. For Kaplan and Ross, Lefebvre's marxism is situated between existential phenomenology and structuralism.[19] Between the existentialists' enthusiastic rediscovery of a Young Marx and the structuralists' urge to delineate a mature, scientific Marx, Lefebvre laid emphasis on the *continuity* of Marx's thought. As he and Norbert Guterman were responsible for some of the first translations of Marx's *The Economic and Philosophical Manuscripts* into French, there is a clear recognition of the way in which the philosophy of praxis, first developed in these early writings, still guides the methodological approach Marx took in his later works. Here there are clear affinities with Gramsci and Lukács, although neither had access to the early writings and, instead, took their lead from fragments such as the *Theses on Feuerbach* and by exploring Marx's critique of Hegel.[20] As Elden notes, Lefebvre posits alienation as the central notion of philosophy and as the basis for a *concrete* humanism, the new humanism he seeks to develop through a refocused philosophy of praxis.[21] Indeed, Lefebvre did much to foreground the importance of alienation as a key marxist concept against the vicious assaults of both official and structural marxisms. Rather than limiting the conceptual purchase of alienation to the *The Economic and Philosophical Manuscripts*, he sees an interest extending throughout Marx's oeuvre. Thus, in the first volume of the *Critique of Everyday Life*, he writes of how "the theory of fetishism demonstrates the economic, everyday basis of the philosophical theories of mystification and alienation."[22] This comes out clearly in his early work on *La Conscience Mystifiee*.[23] Alienation gains a new salience within Lefebvre's *Critique of Everyday Life*. Here it enriches his understanding of the dramatic reduction of everyday life to the repetitions of everydayness. Furthermore, in the theory of the moment, we begin to see the radical potentials Lefebvre discovers in disalienating moves. The moment comes to negate the alienations of everydayness. In contrast to the absence of alienation, the moment comes to assert a presence.[24]

A further—some would argue *the*—crucial aspect to Lefebvre's distinctive marxism is found in his transformation of the dialectic. As with the best of dialecticians, simple binaries are always exceeded: Lefebvre dwells on contradiction, on latent possibilities, on relational thinking and a nuanced sense of totality. Long recognized for this dialecticism, Sartre once commented of Lefebvre that his grasp of the dialectic "is beyond reproach."[25] While the reception of *The Production of Space* has been marked by the recognition that something remarkable takes place in Lefebvre's handling of the dialectic, there has often been disagreement over what exactly it is that Lefebvre does. Thus, Soja makes a claim for a "trialectic" approach that goes on to inform his *Thirdspace*.[26] Shields, too, claims that the dialectic is spatialized in Lefebvre's hands.[27] Elden criticizes both approaches, arguing that "the fact that Lefebvre uses this understanding [of the dialectic] to rethink the question of space . . . does not mean that the dialectic is spatialized."[28] He goes on to argue that one of the more distinctive features of Lefebvre's nonteleological dialectic must be found in the substitution of the Hegelian understanding of *Aufhebung* with the Nietzschean sense of *Überwinden*.

Convincingly, Schmid argues that none of these claims entirely grasps the fundamental transformation of the dialectic Lefebvre performed.[29] To comprehend fully the triadic dialectic Lefebvre developed, we need to understand his engagement with, and critique of, Hegel, Marx, and Nietzsche. Through these thinkers, he establishes three moments: these exist in interaction, in conflict, or in alliance with one another. The first moment Lefebvre takes from Marx. This consists of material social practice. The second, consisting of knowledge, language, and the written word, he takes from Hegel. And the third moment, consisting of *poesy* and desire, Lefebvre takes from Nietzsche. Summarizing these moments in *The Production of Space*, Lefebvre states that the critique of philosophy will necessarily consist of "a confrontation between the most powerful of 'syntheses'—that of Hegel—and its radical critique; this critique is rooted on the one hand in social practice (Marx), and on the other hand in art, poetry, music and drama (Nietzsche)—and rooted, too, in both cases, in the material body."[30]

Lefebvre criticizes the Hegelian dialectic from two perspectives. First, he argues, following Marx, that it needs to be grounded in the everyday rather than the conceptual realm. Second, he goes on to criticize Hegel for "positing a total a priori object—absolute knowledge, the system." In doing this, he "went against the content, against the Becoming, against living subjectivity and negativity."[31] Instead, Lefebvre "posits the metamorphosis of the sign: *poesy*. In his view, the work of art alone is the unity of the finite and the infinite, endlessly determined and living."[32] The three moments Lefebvre introduced are clearly crucial throughout his work. *The Production of Space* is thus constructed around a doubly determined triadic structure. Christian Schmid notes that it was only later in his life that Lefebvre became fully aware of the manner in which he had transformed the dialectic. In one of his more infamous expressions, Lefebvre states: "Wherever the infinite touches the finite there are three dimensions. . . . There are always Three. There is always the Other."[33] As I go on to argue, Lefebvre's transformation of the dialectic, put to work in his analysis of the production of space, is fundamental to how we understand the possibilities for the politicization of nature. Even if his understanding of nature does not provide the active moment for political transformation that one might hope for, we can find traces in his method for developing an alternative conception. From this, it might be possible to develop an immanent critique of the nature of everyday life. First, however, we need to understand how Lefebvre posits the possibilities for change within the realm of the everyday.

The Critique of Everyday Life

> There can be no knowledge of society (as a whole) without critical knowledge of everyday life in its position—in its organization and its privation, in the organization of its privation—at the heart of this society and its history. There can be no knowledge of the everyday without critical knowledge of society (as a whole). Inseparable from practice or praxis, knowledge encompasses an agenda for transformation. To know the everyday is to want to transform it. Thought can only grasp it and define it by applying itself to a project or programme of radical transformation. To study everyday life and to use that study as the guideline for gaining knowledge of modernity is to search for whatever has the potential to be metamorphosed and to follow the decisive stages or moments of this potential metamorphosis through: it is to understand the real by seeing it in terms of what is possible, as an implication of what is possible. For "man will be an everyday being or he will not be at all."[34]

The critique of everyday life is at the heart of Lefebvre's life work and central to his contribution to marxism as a whole. It is utterly exemplary of the possibilities within a marxist philosophy of praxis and the potentials of an immanent critique. As Stefan Kipfer writes, it "was Lefebvre's most enduring concern and thus the linchpin to his conception of marxism as metaphilosophy and critique of political economy."[35] Lefebvre himself saw firm grounds for making the case that Marx's own writings can be read as an attempt to develop a critique of everyday life.[36] Thus, developing the eleventh of Marx's *Theses on Feuerbach*, Lefebvre writes: "We will therefore go as far as to argue that critique of everyday life—radical critique aimed at attaining the radical metamorphosis of everyday life—is alone in taking up the authentic Marxist project again and in continuing it: to supersede philosophy and to fulfil it."[37]

Lefebvre's focus on everyday life delineates where we look to change the world. Although towering over everyday life, the state springs from this ground. Change necessarily must germinate and be propagated within everyday life, even if, in the modern world, this, alone, will be insufficient.[38] Thus Lefebvre defines everyday life as "'what is left over' after all distinct superior, specialized structured activities have been singled out for analysis."[39] In another widely cited metaphor, Lefebvre likens everyday life to "fertile soil" from which spring the flowers and the trees.[40] The presence of the flowers and the trees should not distract us from the fertile soil below. In focusing our attention on the everyday, Lefebvre extends his antistatist critique of society.[41] Repeatedly, he emphasizes both Marx's and Lenin's roles as *anti*statist thinkers themselves. Building on Marx's *Critique of Hegel's Doctrine of the State*, he stresses how the abstraction of the state is derived from the abstract antithesis of public and private in the modern world,[42] something Lefebvre argues to be intimately linked to the reprivatization of life. Rather than investing hope in such concrete abstractions, Lefebvre demonstrates a rich sense of the creative possibilities within the mundane. Thus he writes of how "praxis and poiesis [two concepts fundamental to Lefebvre's metaphilosophy] does not take place in the higher spheres of a society (state, scholarship, 'culture') but in everyday life."[43] Significant parts of the contemporary environmental movement seem to have missed such a critique, instead finding solace in some imagined environmental Leviathan able to impose its benevolent environmental will on society. Lefebvre, in contrast, refocuses our attention on the revolutionary potentials within the everyday as a level within the social totality. Somewhere between the outright dismissal of totality and the counterpoint—a fetishism of the total—Lefebvre finds the critique of everyday life.[44]

If Lefebvre is best known for his development of the critique of everyday life, his position is not dissimilar from that of Gramsci or Lukács. All three seek to transform speculative theorizing into a radical and grounded philosophy of praxis. As Alice Kaplan and Kristin Ross might have it, they seek to raise lived experience to the level of a critical concept and to take that concept as the basis for fundamentally transforming the world.[45] They build on what they see as methodological principles within Marx's transformation of Hegel's dialectical thought and its attempt to *realize* philosophy.[46] Nevertheless, if connections are clear, it is also important to recognize the distinctiveness of Lefebvre's contribution.

In part, John Roberts, who foregrounds Lefebvre's contributions within a much longer tradition encompassing Lukács, Gramsci, Arvatov, Benjamin, Barthes, and Vaneigem, provides such as analysis.[47] What unites these thinkers is a Hegelian marxist emphasis on the role of consciousness in the recognition of human beings of their own alienation and a renewed conception of praxis. How one arrives at this revolutionary consciousness differs in important ways. With Gramsci, Lukács, and Lefebvre all taking aim at forms of economism present within the marxisms of their time, each relies on a series of categories of mediation through which class consciousness is realized. Allying Gramsci with Lefebvre, Roberts notes that "whereas Lukács resolves the problem of mediation through the idealized consciousness of the Party, Lefebvre actually insists on the concrete, contradictory and everyday conditions of mediation."[48]

The crucial category of mediation argued to be lacking in Lukács is found in Lefebvre's distinction between "daily life" (*la vie quotidienne*), everydayness (*la quotidiennete*), and the everyday (*le quotidien*).[49] Thus the everyday becomes the concrete terrain over which revolutionary possibilities might be realized. Conversely, everydayness is, as Roberts argues, "the space of historically unrealized species being. . . . If everydayness designates the homogeneity and repetitiveness of daily life, the 'everyday' represents the space and agency of its transformation and critique."[50] The connections with Gramsci's theorization of common sense are clear: for Gramsci, one of the key revolutionary challenges is to transform the rational kernel within common sense and to be able to organize implicit conceptions of the world into explicit, verbal ones, and then practice. Common sense thus serves as a crucial mediator through which hegemony is both consolidated and contested. However, even if Roberts recognizes that Gramsci's "reflexive emphasis on consciousness and the 'concrete' finds a level of conceptual differentiation that is not paralleled elsewhere in the Western Marxist tradition of the 1930s,"[51] he finds that this results in a loss of dialectical tension between object and subject. Hence he claims that Gramsci produces a similar overinvestment in the world-changing capacities of proletarian consciousness to Lukács, albeit Lukács relies on the Party for this to be realized. If Lukács sees the totalizing effects of capitalist society as reducing consciousness to the status of a thing, Gramsci appears to rely on an innate curiosity and inquisitiveness to be able to go beyond this. Lefebvre, however, seems to escape any such criticisms.

However, even if some of his criticisms of Gramsci and Lukács might be unfair, Roberts is right to argue that everyday life refuses closure for Lefebvre: indeed, it is in Lefebvre's theory of the moment that these radical possibilities appear to coalesce. The moment is "*the attempt to achieve the total realization of a possibility*":[52] it represents the power of presence, as opposed to absence. Developed from Hegel, for whom the moment "designates the major figures of consciousness, each of them is a *moment* in the dialectical ascent of self-consciousness."[53] Lefebvre claims his theory of the moment to be more modest. Each moment is perceived, situated, and distanced. "Moments make a critique—by their actions—of everyday life, and the everyday makes a critique—by its factuality—of paroxysmal moments."[54] Merrifield captures the disruptive power of the moment:

> "It disrupted linear duration, detonated it, dragged time off in a different, contingent direction, toward some unknown staging post. The moment is thus an opportunity to be seized and invented. It is both metaphorical and practical, palpable and impalpable, something intense and absolute, yet fleeting and relative, like sex, like the delirious climax of pure feeling, of pure immediacy, of being there and only there, like the moment of festival or of revolution"[55]

The reference to the moment of festival is important. Throughout the *Critique of Everyday Life*, Lefebvre ponders the role of the festival. The festival is important to his understanding of ruptures, of moments of critique, and of the relationship between nature and society. As Merrifield notes, Lefebvre's understanding of the peasant festival is an archetypal example of the latter's method of looking back in order to look to the future: "He envisages the festival as a special, potentially modern form of Marxist praxis that could erupt on an urban street or in an alienated factory. The festival was a pure spontaneous moment, a popular "safety valve," a catharsis for everyday passions and dreams." Importantly, festivals reconnected people with both nature and human nature.[56] In this,

> festivals contrasted violently with everyday life, *but they were not separate from it*. They were like everyday life, but more intense; and the moments of that life—the practical community, food, the relation with nature—in other words, work—

were reunited, amplified, magnified in the festival. Man, still immersed in an immediate natural life, lived, mimed, sang, danced his relation with nature and the cosmic order as his elementary and cosmic thoughts "represented" it.[57]

Lefebvre and Nature

If the festival represents humanity's immersion in a natural life, it also shows something of Lefebvre's theorization of the relationship between nature and society. In one of his two critiques of Lefebvre's conception of nature, Smith notes the importance of Lefebvre's early and enduring experience with the southern French peasantry and the erosion of peasant life in shaping his understanding of nature.[58] The discussion of the festival captures this well. Elsewhere, whether writing on rhythmanalysis, on space, or on the critique of everyday life, Lefebvre appears to develop a quasi-dialectical conception in which nature and society are interwoven in a process of ongoing struggle. On the one hand, Lefebvre understands "nature" as a differentiated unity comprising the social and the natural. On the other hand, he seems to build on a more romantic conception in which nature is *supplanted* by human activity and especially by the abstractions of capitalist modernity. Nature is rich in political possibilities and yet a real stumbling block to fulfilling those radical possibilities. In what follows, I wish to consider the ways in which some of the more radical methodological moves in Lefebvre's work might be developed through rethinking his conception of nature. In particular, I argue Lefebvre's transformation of the dialectic best exemplified in his work on the production of space, and, more important, his rich model of cultural praxis afford fecund starting points for understanding nature than are available in his explicit writings on nature, fascinating though these are. Above all, I ask what it might mean to apply some of the insights Lefebvre develops in *The Critique of Everyday Life* and in *The Production of Space* to questions of the production and urbanization of nature.[59] It is important to note that this is a move that Lefebvre would not have pursued: in spite of his criticisms of such an approach, he consistently develops a twofold conceptualization of nature that, although related to society, is also a realm untainted by human activity.[60]

Put most succinctly, nature, for Lefebvre, is that from which humanity emerges, that which it struggles against, controls, and then seeks reimmersion within.[61] Above all, nature designates both human nature (an emerging process) and the origin from which this human nature emerges. "Human nature," following Marx, is defined relationally and understood as only ever existing dialectically, "in the endless conflict between nature and man."[62]

Lefebvre, nevertheless, seems to revel in the space between these two determinations of nature. Thus:

> The moment human hands, or eyes, or tools, touch pure nature, it is no longer "pure nature." Yet that is how it "is": the open sea, the space between the stars. The contradiction (and thus the double determination) is already apparent in action and in theory, in concept and in practices. It is extraordinarily stimulating. It is forever being resolved in the transition from the "thing-in-itself" to the "thing-for-us," and is forever reappearing at the very heart of this difference.[63]

As this passage also makes clear, if there *is* a dialectical relationship between humans and nature, this is, at least in part, an external, albeit mutually constitutive relationship: it is a relationship of struggle. Thus Lefebvre argues that the only way to grasp the relationship between humans and nature is one of "a dialectical conflict."[64] Within this, the intricacies, levels, and mediations of Lefebvre's three-dimensional dialectic are apparently suspended. Indeed, rather like the Frankfurt School conception, best developed by Alfred Schmidt, Lefebvre's understanding is rather more Kantian than we might expect.[65]

In the ninth prelude to the *Introduction to Modernity,* Lefebvre dwells for some time on the bold claim made by Marx that "industry is the real historical relationship of nature to man." In this, Marx seems to open the possibility for the "production of nature" thesis developed by Smith.[66] Lefebvre, similarly, calls on us to rediscover meaning in this idea and to reconsider the rich meditations on nature found in the Paris Manuscripts. Without this, he argues, the concept of nature will continue oscillating between an idealist ontological interpretation and a materialist one. Returning to a more nuanced dialectical position, he argues that labor, industry, and technology might be seen to be mediations between humans and the nature that they control. Later in this remarkable passage, Lefebvre develops one of the more evocative pictures of this conception of nature when he describes the historic shifts in consciousness of the coming of spring. Nature, he argues is *physis,* a fundamental power and a cyclical becoming. Spring represents one of the clearest examples of its repetitions, one showing that such repetition can never be reduced to mere reiteration. In ancient times, the return of spring represented a period of festival, a celebration of fertility and spontaneity in which "the citizens throw off morality, politics, knowledge, discourse, rhetoric and the games of reason to plunge themselves into the rhythms of the life force."[67] A profound shift takes place in the Middle Ages as this ancient Dionysian springtime gives way to a more solemn cult of the Virgin. Spring is no longer a festival of lovemaking but a "habitual time of purity and virginity." With the Renaissance, a secular reinstatement of spring occurs in which the Christian traditions of the intervening period give way once more to the themes of Antiquity. Lefebvre ponders these shifts, arguing for an explanation that is not merely conducted at the conceptual level (the progressive ideas of the Renaissance challenge the conservatism of the Middle Ages with the themes of the Classical period), but that is rooted in praxis. His explanation is that spring is normally a period of dearth in the agricultural calendar. Situated halfway between the winter festivities of Christmas and the autumn celebrations of Harvest, it marks a time of need, abstinence, and waiting. However, following the partial subordination of such cycles to revolutions in agricultural techniques, reserves can be set aside and the future prepared for:

> And now men—the most "cultivated" men at first, people from the towns and then the masses—rediscover the spring. They are amazed by it. They rediscover nature, long forgotten by their ancestors and their fathers. But this springtime is no longer the springtime which breaks the laws of the city. It is a springtime which has already been controlled and appropriated. The life of nature no longer unfolds before their eyes, something beyond them, an absurd and ludicrous spectacle, its exuberant blossoms threatening death, a dangerous, turbulent elemental disorder, a wild bestial frenzy. At the same time as it resumes its place in the cycle of nature, spring, though still ruled by the law of cycles—becomes subsumed in the cycle of social living.[68]

Within this analysis, nature and praxis cannot entirely be divorced. While nature remains a realm shaped by its own cyclical laws, these laws are increasingly dominated by human social history. Still, whereas humans can appropriate and learn from the cyclical laws, these laws become less fundamental in shaping what humans can and cannot do, forming more of a static backdrop than a force that shapes the lives of those who encounter it. Nature is at once both curiously passive—the struggle to control is central to the emergence of civilization—and yet, potentially, active—shaping, if only in part, the cycle of social living. Elsewhere, Lefebvre summarizes his position: "While it is true that human action impacts back upon the *physis* it issues from, it is in order to unfurl it in a second, infinitely rich and complex nature—products and works—but at the risk of destroying the first nature and severing the increasingly frail nutritive bond that links the two (we might say, with Spinoza; *natura naturans* and *natura naturandum*)."[69] He continues: "If there is a reconciliation, or at least a compromise, between first and second natures, it will occur not in the name of an anthropological or historical positive knowledge, but in and through daily life."[70] Again, in both passages, Lefebvre employs two determinations of nature—one more closely associated with our understanding of human nature and captured in a sense of emergence; the other in the sense of origin, that from which this human nature must emerge. Humans struggle to free themselves from the realm of necessity implied by their total submersion in Nature as origin: in the process, they shape a new, second social or human nature. Nevertheless, reconciliation, differing from the first nature that existed previously, is to be sought, and this must occur at the level of everyday life.

Similar distinctions—between a first nature devoid of human activity and a second nature shaped by praxis—occur throughout Lefebvre's writing. Stuart Elden, for example, notes the distinction (influenced in part by Lefebvre's relationship with Axelos) between the earth—*le terre* and the world—*le monde*: "the earth, the Planet Earth, becomes the world through our intervention."[71] In the crucial distinction Lefebvre makes throughout *The Critique of Everyday Life* between need and desire, his two determinations of nature reemerge. Need is seen as part of first nature, something from which, in the struggle to feed, clothe, and harbor humanity from external danger, humans have freed themselves and replaced with "desire."[72] Again these two determinations of nature would appear to come out in Lefebvre's lifelong musings on time. Cyclic time, as in the example of the coming of spring, is rooted in a primordial nature. Nothing separates nature from social life in cyclic time, but though persisting in the background this bond is ripped apart by the abstractions of capitalism and the linear time of modern industrial society. Thus: "Cyclic time is replaced by a linear time which can always be reckoned along a trajectory or distance.... *Critique of everyday life studies the persistence of rhythmic timescales within the linear time of modern industrial society. It studies the interactions between cyclic time* (natural, in a sense irrational, and still concrete) *and linear time* (acquired, rational, and in a sense abstract and antinatural)."[73] This of course was the project Lefebvre was later to deepen in *Rhythmanalysis*.[74]

For Smith, these two understandings of nature in Lefebvre lead to a fundamentally antinomian conception of space and nature. Whereas Lefebvre makes the radical move of investing *space* with agency, he fails to extend such an analysis to nature. Instead, as Smith argues, in Lefebvre's somewhat teleological analysis, nature is increasingly emptied of its distinctive content through the development of capitalist modernity. Smith constructs this argument by focusing solely on *The Production of Space*. Within this, he demonstrates how

> nature for Lefebvre is on the verge of becoming a corpse at the behest of abstract space. Unlike space, nature retains virtually no initiative to and for itself. "Displaced and supplanted" by social space (123, 349), it is destroyed on the one

hand, and at best "merely reproduced" on the other (376). It is space that progresses and develops, is produced and conjures, space that is alive. Space can even lie, Lefebvre contends, but nature cannot: "A rock on a mountainside, a cloud, a blue sky, a bird on a tree—none of these of course can be said to lie" (81).[75]

And later:

> This bifurcated treatment of space and nature emanates ultimately from Lefebvre's handling of production. Production, he argues, is a quintessentially social act, and while nature creates and thereby provides use values for production, it does not produce (68–70). Space and commodities, however, are social products; they lie and social space itself can produce. Now this gives a surprisingly undialectical negativity to the treatment of nature, especially vis-à-vis the life accorded to space. Where the dialectic of space is defined by its political vitality—social and political practice *is* spatial practice—the dialectic of nature is evacuated. Nature is reduced to little more than a substratum.[76]

Unsurprisingly, Smith advocates a radical reinterpretation of nature as that which is produced (this was discussed further in chapter 1), a step that as Smith shows, Lefebvre seems ready to accept—"There is nothing, in history or in society, which does not have to be achieved and produced. 'Nature' itself, as apprehended in social life by the sense organs, has been modified and therefore, in a sense produced"[77]—only then to go on to dismiss such a claim. If Smith's criticisms of Lefebvre's antinomian reading of space and nature are to a large extent true for *The Production of Space*, they are perhaps too harsh for the understandings developed elsewhere in Lefebvre's work. Here, at least, there is a sense that Industry does indeed produce a differentiated unity or socio-nature. Nevertheless, this remains a differentiated unity in which Nature, as a separate primordial realm, is increasingly inert.

In part, following Smith, such antinomies can indeed be overcome by focusing on the production of nature. However, to get to this position I would like to follow some of the steps within Lefebvre's own way of working: this leads to a method that is compatible with positions Lefebvre holds elsewhere. More specifically, I will argue that Lefebvre's model of cultural praxis holds out hope for a renewed understanding of the nature of everyday life that draws on the best of Lefebvre while countering his antinomian approach to nature. First, however, I will turn to Lefebvre's model of cultural praxis.

Lefebvre's Model of Cultural Praxis

> If it is true that culture can no longer be conceived outside the everyday, this is also true of philosophy and art, which constitute the center of the cultural and

define its axes. For many Marxists, it seems that art is only a distraction, a form of entertainment, at best a superstructural form or a simple means of political efficacy. It is necessary to remind these people that great works of art deeply touch, even disturb, the roots of human existence. The highest mission of art is not simply to express, even less to reflect, the real, nor to substitute fictions for it. These functions are reductive; while they may be part of the function of art, they do not define its highest level. The highest mission of art is to *metamorphose* the real. Practical actions, including techniques, modify the everyday; the artwork transfigures it.[78]

Even if he is critical in the first volume of the *Critique of Everyday Life* of the work of the Surrealists, claiming that "the Surrealists promised a new world, but they merely delivered 'mysteries of Paris'"—Lefebvre clearly drew great sustenance from both aesthetic and literary forms of cultural practice. Thus, one finds repeated references to *Ulysses*—for the "momentous eruption of everyday life into literature."[79] Elsewhere, as with Benjamin, he finds inspiration in Brecht's Epic Theater, which "immerses itself in everyday life, at the level of everyday life."[80] "Literature and art," Merrifield writes, "have better grappled with understanding the everyday."[81] Beyond this, Lefebvre finds inspiration for an artistic mode of working in Marx, who, as the former reminds us on several occasions, possesses both an ethical and aesthetic critique of contemporary society. This aesthetic critique is of crucial importance for Lefebvre:

> The creative activity of art and the work of art foreshadow joy at its highest. For Marx, enjoyment of the world is not limited to consumption of material goods, no matter how refined, or to the consumption of cultural goods, no matter how subtle. It is much more than that. He does not imagine a world in which all men would be surrounded by works of art, not even a society where everyone would be painters, poets or musicians. Those would still only be transitional stages. He imagines a society in which everyone would rediscover the spontaneity of natural life and its initial creative drive, and perceive the world through the eyes of an artist, enjoy the sensuous through the eyes of a painter, the ears of a musician and the language of a poet. Once superseded, art would be reabsorbed into an everyday which had been metamorphosed by its fusion with what had hitherto been kept external to it.[82]

For Feenberg, such a society depends on a fundamentally different apprehension, and thus production of nature through sensuous human activity. Thus: "Revolution unites subject and object in liberated sensation and thereby reveals the truth of nature."[83] However, we might also take a different road in which, following Smith, the production of nature is understood in a more radical sense: nature, too, as I have argued earlier, *is* separated from society in late capitalism, but this separation emerges from historically and geographically specific mediations that produce nature in all its immediacy as separate.

If this argument appears contradictory, one need only look at the examples from Durban that I developed earlier. Water is produced as a socio-natural entity that embodies and expresses the relationships of capitalist society. In the process, it exerts a power over residents of informal settlements and townships throughout the city. Democratic access to nature, as I have argued, might be pursued through an immanent theorization of the abstract production of nature under capitalism. This is exactly the move Lefebvre made in his understanding of the creative potentials within the culture of everyday life. Indeed, if art appears within Lefebvre's writings as a separate domain, this is precisely because of the mystifications and alienations of neocapitalism. In the closing pages of *Everyday Life in the Modern World,* he calls for us to "Let everyday life become a work of art!"[84] This rallying cry counters the splintering effects of modern industrialism: it is a plea that we extend artistic modes of working into the production of everyday life. As Roberts argues: "The production of culture lies in the reconquest and immanent theorization of alienated, industrialized experience," leading to "a quasi-Productivist notion of art as social transformation" in which art *is* praxis:[85]

> The outcome is a theory of culture that places a primary emphasis on the extension of the form of the artwork and aesthetic experience into the environmental and architectural. Following the lead of Surrealism, the forms and symbols of the city are incorporated into an expanded field of reference for art and, simultaneously, the sites of the city become the actual locations of artistic practice. The city is taken to be both a work of art requiring interpretation (a place where meanings are generated) and a place where art is practically realized (a place where the context of the work's staging becomes part of the work's or event's identity).[86]

It is this dimension of Lefebvre's writing that has been picked up on in recent work on urban interventions and critical spatial practices. In his extension of the philosophy of praxis to artistic production and in his understanding of the possibilities immanent in the acts of producing space and art in the contemporary moment, Lefebvre provides hope to a range of both theorists and practitioners.[87] Within this, there is a clear recognition that Lefebvre's emphasis lies as much in creating "extra-institutional events as significant forms of cultural intervention, as it is a theory of aesthetics."[88]

Nevertheless, Lefebvre's work has had almost no impact on reinterpreting the politics of the environment. And yet remarkably similar conceptual moves are made within work on the cyborg city that may well make a conversation possible between critical spatial practices and urban political ecology. For within recent theorizations of metropolitan nature, the city is shown to be a field constituted out of socio-natural processes: it emerges from the ongoing coevolution of nature and society. In recognizing how the city is produced, reproduced, and transformed by socio-natural processes, praxis is extended to what is, at least potentially, an urban environmental politics. This, to reiterate, is a politics rooted in quotidian relationships and in which consciousness of the socio-natural is immanent to these sensuous practices. It differs profoundly from dominant environmental movements in which an environmental vanguard is responsible for instilling new ways of working on an otherwise unthinking populace. If we extend the boundaries of praxis into both the socio-natural realm (it was always there, although not explicitly) *and* extend these boundaries into the aesthetic experience, recapturing sensuous experience in the process, then, I argue, we have the possibility for a radical politics from which we might remake our cities in sensuously rich, radically democratic and beautiful ways. With this, I would like to return to the example with which I began this chapter, City Mine(d)'s strange Ping Pong Project in the cul-de-sacs of Wembley Park.

City Mine(d)'s Urban Praxis

As I alluded to in the introduction to this chapter, the Ping Pong Project consisted of constructing a network of plastic tubing linking some of the key community facilities in an area of Brent, a suburban London borough in the north of the city, in early 2006. On a cold March afternoon, the roads around Preston Road tube station, where the network was assembled, felt far from the typical artists' playground of Hoxton or Brick Lane: more like the streets of Mourenx that Lefebvre so despised, the area felt managed, regulated, and reigned by everydayness. With the network assembled, a hydraulic pump was installed, meaning that ping-pong balls could be fired for the length of the network: local residents were invited to write thoughts, inspirations, protests, or musings onto these balls before inserting them in the network. New conversations might be stimulated in the process. Prior to this, the process of constructing the network involved bringing together new coalitions of actors, charting incredibly complex webs of planning regulations, and provoking a range of debates about the role of art, the disconnected "community," and the use of public and private space.

With the Ping Pong Project, as with other interventions, City Mine(d) succeeded in turning the ecology of this neighborhood into a laboratory for radical experimentation. From this, conditions of possibility are generated for new ways of relating and the socio-natural makeup of the city might thereby be reimagined. In this process, the apparent primacy of the artist as the creative subject and the finished work as the object of analysis is challenged. There are obvious echoes of Nicolas Bourriaud's claims around the development of a relational aesthetics in the 1990s:[89] here the artist is positioned within a matrix of relations and the finished work of art becomes less important than the artistic act. Thus, even though they would no doubt reject their positioning within such an artistic turn, with the Ping Pong Project, City Mine(d) fostered a participatory matrix or "relational aesthetic" that operates through the physical network of tubing and ping-pong balls. On completion, the messages, the communication, and the very success or failure of the work are decided upon by the participants themselves: these participants embark on reworking the socio-natural makeup of their own community. Here the project was always much more about relating to urban processes than finished products: it was about metamorphosing the nature of everydayness as it existed in Brent Borough (in a "chemically pure state") into the act of artistic production. An artist who works closely with City Mine(d), Chris, described this process-ethic in what are almost Dadaist terms: "When we came up with the idea of the ping-pong network, we spent a long time trying to think about what we should do with the balls once they'd been collected. After hours of thinking about it, we realized we should do nothing. Absolutely nothing. The balls are fired and that is it."[90] Funded in part by the local London borough, this made explaining the project's outcomes problematic and opened both fissures and disagreements about the role of artists and the political relations traversing the borough.[91] The mayor's appropriation of the project, as a means of communicating messages of concern to the local council, neatly sidestepped the embarrassment of having funded such Dadaist lunacy. At the same time, it brilliantly failed to capture what the project was seeking to, or, in reality, did achieve. However, neither the anarchy of Dada nor mayoral normative aspirations actually captured the conditions of possibility of the project.

Instead, the project generated conditions of possibility for new forms of consciousness of metropolitan nature. Through expanding creative praxis into this socio-natural realm, it demonstrated how everyday metropolitan nature is actively constituted out of a range of relationships. At the same time, it provided opportunities to remake those relationships: in providing creative entry points, again both physically—as people had an open tube in their local pub, and socially, as the project opened up a new dialogue between key local agents, the Ping Pong Project turned a relational and process-based understanding of the city into an infrastructure for transcending cultural praxis and everyday natures.

However, to argue that the acts of City Mine(d) are Lefebvrean, Gramscian, or political-ecological is perhaps to miss the point. Rather, the form of cultural praxis they develop provides insights into how we might take forward the form of critique I have advanced in this book. They develop a praxis that has affinities with the projects advanced by each of these theoretical frameworks, even if their acts are not informed by any of them.

Conclusions

City Mine(d)'s mode of working specifically informs the reading of Lefebvre developed in this chapter. First, I argued that one of Lefebvre's signal contributions to marxist social theory is to develop a philosophy of praxis rooted squarely within the everyday. This is central to the way City Mine(d) works. The collective has set up sofas on roundabouts, sent gigantic plastic balls bouncing through neighborhoods in Brussels, worked in schools and in run-down, drug-infested alleyways. City Mine(d)'s practical work is tied to those spaces that remain when one removes all the elite spaces of authority and decision making. It is manifestly everyday. If Lefebvre is right to call for us to "Make everyday life a work of art!" this is, as Roberts notes, more than a call to make art public. The development of public art in recent years has been criticized from a number of perspectives. It feeds the real estate market, bolsters the interests of nations and states, and reasserts the divide between the artist and the community with which they work. (Antony Gormley's depressingly dull *One and Other* on Trafalgar Square's Fourth Plinth in London served as one of the more recent and higher profile reminders of this.) Lefebvre's critique is more radical. More radical too than suggesting that we should all be artists. Instead it is to argue that the creative act of making the artistic object should be extended to the making of life itself. In the case of the socio-natural makeup of the city, it is to argue that, sensitively and humbly, we might embark on a creative process of remaking urban natures in fundamentally more democratic and ecologically progressive ways.

This is the second important point. Lefebvre, I argued, has a rich model of cultural praxis, he reworks the dialectic in incredibly suggestive ways and he is able to apply this understanding to the production of space; however, he does not provide much ground for us to build an ecological politics. Although never explicitly, City Mine(d)'s interventions show how a cultural praxis metamorphosed into the everyday could also be a socio-natural politics. There are strong grounds for demonstrating that Lefebvre's model of cultural praxis could be applied to an understanding of the socioecological makeup of metropolitan nature. Let's make this move and, in the process, turn a critique of everyday life into a critique of metropolitan nature by expanding the philosophy of praxis, the dialectical method, and the understanding of cultural praxis that Lefebvre develops into the ecological. In the course of this book I have argued for the same—and there are firm grounds for doing so—in Lukács, Gramsci, and Marx.

The third and final point is crucial. Lefebvre, as with the other thinkers I have considered, has a rich immanent critique of everyday life. Developing a radically democratic environmental politics for the present moment, I argue, depends on an exploration of the new forms of consciousness that might be possible in everyday socio-natures. This consciousness will not be imparted by some external agent acting as the Party might have acted in days gone by. Nor will it arise spontaneously from the sudden realization of impending doom. Nor should it be essentialized as the end point for some already imagined coherent subject. Instead, I argue, a radical ecological consciousness is likely to develop in the everyday through practical sensuous activity. In the moments of rupture that punctuate the encroaching rationalizations of everydayness, we might forge new forms of solidarity with the human and non-human world. Lefebvre guides us toward these forms, just as City Mine(d), with their ping-pong lunacy, did in Brent.

Conclusion The Nature of Everyday Life

> Dessous les pavés la plage (Beneath the cobblestones lies the beach)
> —Maxim circulated widely during the uprisings of
> May 1968 in Paris

IN THE COURSE OF WRITING THIS BOOK, the world known to many will have been dramatically reconfigured. Trumping my own hopes for a changed world, capital has achieved its own perverse transformation. Aided by willing political handmaidens, and in depressingly familiar ways, the outcome of the latest economic crisis has been a world manufactured even closer to the desires of finance capital. One need barely look any further than the university in which I work as it is disassembled and reassembled according to the needs of business. Or perhaps one might look at the city in which I live, where a fusion of speculative capital and attacks on the welfare state are permitting forced removals of those on low incomes. Or perhaps the play parks, the libraries and the museums, all of which now sit under the Damoclean sword of a neoconservative administration carrying out what it claims is demanded by international bond markets.

Elsewhere in the world, capital has simply packed up and moved on, leaving the cities that were manufactured for it empty apart from the ghosts of industrial ruins. The stifled anger I wrote about in the introduction has grown and found expression in a variety of contradictory movements. Populist mayors have been elected in many cities of the global North, and the Tea Party movement in the United States has captured the world's attention: many, it would seem, have sought solace in the promises of monsters whose offers to assuage fears of this unfixable world seem to provide some kind of a balm.

Nevertheless, the present moment might also be one of opportunity. There are more obvious examples in the incipient student movement that was born in the UK. Formed in opposition to the dismantling of a university system, it has rediscovered a world of desire, imagination, and freedom. This is a movement that rages against the austere, miserable world that capital has crafted. Most likely it will not be "the beginning of the end," as Quattrocchi and Nairn claimed of the Parisian uprisings in 1968. But it has opened a world of revolutionary yearning that will be hard to close down: "When the unpredictable has happened, the seemingly impossible is at hand."[1]

But there are less obvious examples that perhaps run deeper. Outside the world of the revolutionary dreamer, the economic crisis provided one of the clearest demonstrations of the mutability of the contemporary world, an argument that lies at the heart of this book. There was something remarkable in those moments when flows of capital began to stall. In every shop, bar, and play park of this city, people recognized that the world they knew, loved, hated, or simply took for granted was held together in such fragile ways. Of course capitalist society did not fracture. Nor did its architects even dwell on this fragility for long. Instead, elites fought hard to close down such questions and transform what was a clear legitimacy crisis into a public-sector debt crisis that appeared to require an attack on the weakest. But we did question, we really did.

In what appeared to be the dark days of the 1990s (the days have only gotten darker), Donna Haraway admitted that she found it harder and harder to imagine a world beyond capitalism.[2] And yet, there we were, in the florist at the end of the road, in the children's play park on the other side, daring to imagine. I was drawn to a moment in the novel *Germinal* when the hero, Étienne Lantier, arrives in the town of Montsou looking for work. He inquires of the owner of the mine in the hope of securing employment there. The inverted world we live in is captured brilliantly as Lantier questions one man whose voice takes "on a kind of religious awe; it was as if he [Lantier] had spoken of some untouchable tabernacle which concealed the crouching greedy god to whom they all offered up their flesh, but whom they had never seen." In the moments around the autumn of 2008, the crouching greedy god somehow looked a little more sick than he (and I'm sure he is a *he*) had before. Indeed, he seemed less a god and more like the wizard of Oz, hiding his frailty behind an intricate disguise. Lukács recognized the importance of moments like this and his immanent critique of everyday life is structured around workers' abilities to make sense of them. And this applies not only in moments of crisis but within the daily acts of making and remaking the world. Feminist standpoint theorists radicalized such a conception, building on the conditions of possibility within and ridding Lukács's analysis of its backdoor essentialisms. In the process, they drew attention to the socio-natural relations that lie at the heart of such an understanding. Urban political ecologists, in turn, have sought to do the same within the city. Here, beneath the cobblestones, lies the beach.

I mentioned this *soixant-huitard* aphorism to my four-year-old daughter this morning. Her scooter rattled over the grit and cobblestones, frustrating her journey to school. Trying to make some sense of what the revolutionary dreamers of 1968 might have meant, I spoke of a world in which our lives would not be structured by the constraints of cobbles, or by work and institutions, but rather by the freedom to participate fully in the making of the city. This would be a joyful process. And I spoke, too, about the ways in which our cities conceal connections with one another and with what lies beneath. We both like beaches, so I think it made sense on one level. But I also think Rosa could detect the political ecological implications—she loves to think of a world of water pipes, cables, and sewerage down below. And if cities can be made out of sand or cobblestones, Rosa knows there is always another world that we don't get to see. The choreographies of power that give shape to cities in their current form might be dominated by those rich enough or invested with enough authority to be able to do so. But that need not always be the case. Beneath the thing we understand to be the city are processes that momentarily allow it to appear as a specific thing. Making, remaking, and unmaking this world—whether one in the image of capital, or one to be crafted by democratic involvement in the process of production—is a process of mutual coevolution between human and nonhuman. Urban environments are one of the concrete manifestations of this process of mutual coevolution.

In an example he returns to on several occasions, Erik Swyngedouw writes of the ways in which this process of coevolution has come to internalize the dynamics of a crisis-prone system. In this way the social ecology of a city like Jakarta was transformed by the implosion of the Southeast Asian financial bubble of the 1990s and El Niño's global dynamics. The convergence of the twain permitted malaria and dengue fever to reappear among the silent construction sites and discarded labor force.[3] If the financial bubble of the 1990s provided evidence of this, it has only been reinforced in the latest crisis. But as Swyngedouw's example makes clear, thinking in this way means thinking in radically different ways about what makes for an environmental politics. The environmental influence of the latest financial crisis has been felt in a vast array of "socio-natural metabolisms," from the partial collapse of the European Union Emissions Trading Scheme to the potholed roads of London. In this regard, the failures of mainstream environmental thinking have, I would argue, only become more stark.

When I first turned to write this conclusion, snow was falling on the streets of London. In the shops, bars, and barbershops of my neighborhood, people muttered about the myths of global warming and the UK was gripped by one of the coldest winters in years. Only a couple of months earlier, world leaders had failed to strike a deal that would have forced them to limit carbon emissions, following a summit that could not have been a better demonstration of the mess of contemporary environmental politics. On top of this, a series of leaked e-mails from the University of East Anglia embarrassed the scientific community and gave succor to the crudest skepticism of climate change. BBC radio phone-ins were filled with theories about conspiratorial bands of scientists somehow seeking to control global consumption habits with their cultish practices. Meanwhile, the head of the International Panel on Climate Change was made to look foolish and weak for his failure to apologize for relatively insignificant errors in the panel's most recent report. This backlash against environmental politics has since gained steam—only to be overtaken by fears of impending bankruptcy that the Far Right has cynically managed to level not at the financial markets but at the state provision of public goods. By pitching environmental arguments at grand levels of abstraction and by forecasting apocalypse, they have cultivated a sense of disempowerment and, importantly, a sense of fear. While distrust for figures of authority is surely necessary for liberatory change, a transformative politics cannot be built on disempowerment and fear. Both serve to bury that different world further beneath the cobblestones, the asphalt, or the concrete.

This book has therefore been an attempt to demonstrate the centrality of nature–society relations to the making of everyday life—within the scoot to school, the gendered division of labor, or the act of ensuring that a household has sufficient water to survive. It seeks a different politics within the hubbub of environmental backlash, revanchist financialization, and statist failures and at the same time builds on a frustration with the high politics of global summits. The mutability of the contemporary world, shown by the financial crisis, is a help in this regard: the challenge for those of us seeking an anticapitalist politics is to transform relationships in ways that foster a world where all can participate democratically in its production.

Thus I have constructed the book around three related arguments. First, it is concerned with the way in which the world is made. This, I have argued, should be seen as a process of mutual coevolution between human and nonhuman. Changing the ways in which humans relate to one another will have an influence over the ways in which nonhuman and human interrelate. But against the hubris of technologically or economically determinist arguments, what we once conceived of as "nature" is an active agent in this crafting of future worlds. In the last three hundred years, the rise of a capitalist system of accumulation has served to utterly transform the ways in which humans relate to this socio-natural world. Within the global North, the move from a feudal system in which peasant farmers worked the land directly, to one in which commodities are produced and exchanged for the acquisition of profits, has utterly transformed the way in which these moments are woven together. For this reason, I have sought to build on both Marx's understanding of this metabolic process and Neil Smith's brilliant excavation of the production of nature.

Second, the book is concerned with the ways in which different people might struggle to make sense of the production of this world. Within the quotidian acts of relating to one another and to "nature," there are conditions of possibility for conceiving life differently. If nature is, in part, produced through productive and reproductive acts, consciousness also emerges in this process. Whether through caring for a loved one, working in a call center, gathering water, crafting a sculpture, or building a boat, practical activity is one key moment in the formulation of worldviews. Rather than seeing this process in abstract terms, I argued that the making of environments is part of a deeply sensuous process. The metabolic process that Marx describes—and that lies at the heart of both the production of nature and Lukács's faith in the working class's dialectical viewpoint—is experienced through sensuous interactions. But this sensing in common can generate quite different worldviews, and we should be wary of romanticizing some already-formed-and-only-to-be-grasped standpoint of a coherent proletariat. Antonio Gramsci, looking back on the reversal of a revolutionary moment, was only too aware of the contradictory nature of *senso commune*. For him, the key to a philosophy of praxis is the ability to transform this common sense, building an immanent critique from within but producing something unrecognizable in the process and fit for the construction of the new civilization in which he had so much faith. Thus, the book is about the contradictory ways in which people make sense of their role in producing the world—a role that is always a process of mutual coevolution in which nature is a partner.

Third, if we recognize our role within this coevolutionary process, we open up new potentials for remaking the world in ways not constrained by the dictates of capital. Allying this recognition with a model of cultural praxis and bringing to light the creative processes involved in making and remaking the world provides further possibilities. In their foreword, Quattrocchi and Nairn write of how "the golden moon in the events of May [1968] . . . cannot be seen with the eyes, only with poetry and the most abstract of thoughts."[4] The events themselves were a poetic attempt to show how the city might be constituted differently. Lefebvre of course recognized this when he argued that "art can become *praxis* and *poiesis* on a social scale: the art of living in the city as a work of art,"[5] and Merrifield concurs, writing that "politics more than anything needs the magical touch of dream and desire, needs the shock of the poetic."[6] Without this poetry the truly liberatory potentials within such a critique are lost and the alternative worlds we might wish to construct will merely echo the past. We need to draw our poetry from the future and in the process rediscover the creative spirit of revolution.

So this book is both a critique of the world in its existing form and a search for future possibilities: the future possibilities emerge from within this process of critique. It lays out both a field of study and a terrain for action in "the nature of everyday life." Of course there is no essential nature of everyday life, but there are activities involved in the creation of everyday natures that might form the basis for an immanent critique. And there is no blueprint for how to proceed, for to produce such a misguided plan would be to deaden such possibilities. Each chapter has therefore explored how our world is constituted, the tendencies within the changing constitution of the world, and the principles that might be built upon in seeking change. I remain utterly convinced of the possibilities within a collective project to make a better world. The time has come, however, to close this book and to start creating those future ecologies that lie beneath the frustrating cobblestones.

Notes

Introduction

1. Donna Haraway, *When Species Meet* (Minneapolis: University of Minnesota Press, 2008), 3.

2. Lucien Goldmann, "Reflections on History and Class Consciousness," in *Aspects of History and Class Consciousness*, ed. István Mészáros (London: Routledge & Kegan Paul, 1971), remarks that the rebirth of dialectical thought, as with Marx's own transformation of Hegel, corresponds with certain social realities: the turn to Hegel by Lenin, Lukács, and Gramsci corresponds to an age of revolutionary upheaval (66).

3. Karl Marx, *Capital: Volume 1* (London: Penguin, 1976), 103.

4. For a classic periodization of art's changing role in relation to broader social forces, see Peter Bürger, *Theory of the Avant-Garde* (Minneapolis: University of Minnesota Press, 1984).

5. Jacques Rancière, *The Emancipated Spectator* (London: Verso, 2008), has, in part, argued against such an understanding, claiming that this distance is assumed by critical theorists and contemporary artists and that, in contrast, the spectator has always been active and engaged. However, in the case of the physical separation referred to here, I see no problem in referring to this distancing.

6. Jane Rendell, *Art and Architecture: A Place Between* (London: I. B. Tauris, 2006).

7. Henri Lefebvre, *Everyday Life in the Modern World* (London: Continuum, 2002), 204; Henri Lefebvre, *Writings on Cities* (Oxford: Blackwell, 1996), 147.

8. Marx, *Capital: Volume 1*, 279; Karl Marx, *Early Writings* (London: Pelican, 1975), 348–50.

9. Harold Wolpe, "Capitalism and Cheap Labour-Power in South Africa: From Segregation to Apartheid," *Economy and Society* 4, no. 2 (1972): 425.

10. Jared Diamond, *Guns, Germs, and Steel: A Short History of Everybody for the Last 13,000 Years* (London: Vintage, 1997). For critiques, see Andrew Sluyter, "Neo-Environmental Determinism, Intellectual Damage Control, and Nature/Society Science," *Antipode* 35, no. 4 (September 2003): 813–17; and (perhaps surprisingly) for a more forgiving critique, see David Harvey, *Cosmopolitanism and the Geographies of Freedom* (New York: Columbia University Press, 2009).

11. Jeffrey Sachs, *The End of Poverty: Economic Possibilities for Our Time* (London: Penguin, 2005); Richard Peet and Elaine Hartwick, *Theories of Development* (London: Guilford Press, 2009), 135.

12. John Bellamy Foster, *Marx's Ecology: Materialism and Nature* (New York: Monthly Review Press, 2000).

13. Rachel Carson, *Silent Spring* (London: Hamish Hamilton, 1963).

14. For one example, see Ted Benton, "Marxism and Natural Limits: An Ecological Critique and Reconstruction," *New Left Review* 1, no. 178 (November–December 1989). The Romantic tradition remains important for much of Benton's work on red–green politics and his work in the Red–Green Study Group.

15. Stephen Daniels and Georgina H. Endfield, "Introduction: Narratives of Climate Change," *Journal of Historical Geography* 35, no. 2 (2009): 215–22.

16. Paul Kingsnorth, "A Windfarm Is Not the Answer," *The Guardian*, July 31, 2009.

17. Erik Swyngedouw, "Impossible Sustainability and the Postpolitical Condition," in *The Sustainable Development Paradox: Urban Political Economy in the United States and Europe*, ed. Rob Krueger and David Gibbs (London: Guilford Press, 2008).

18. John Roemer, *Analytical Marxism* (Cambridge: Cambridge University Press, 1986), 191.

19. It could also be noted that if Bertell Ollman can create a dance of the dialectic (open to anyone with a vague sense of rhythm), this claim of obscurantism seems flawed.

20. Bertell Ollman, *Alienation: Marx's Concept of Man in Capitalist Society* (Cambridge: Cambridge University Press, 1973).

21. Ibid.

22. David Harvey, *Justice, Nature, and the Geography of Difference* (Oxford: Blackwell, 1996). Perhaps an even clearer contribution can be found in Harvey, *Cosmopolitanism and the Geographies of Freedom*.

23. Nancy C. M. Hartsock, "Moments, Margins, and Agency," *Annals of the Association of American Geographers* 88, no. 4 (1998): 708.

24. David Harvey, "On the Deep Relevance of a Certain Footnote in Marx's Capital," *Human Geography* 1, no. 2 (2008): 26–31.

25. Marx, *Capital: Volume 1*, 493.

26. Harvey, "On the Deep Relevance of a Certain Footnote in Marx's Capital," 29.

27. Hartsock, "Moments, Margins, and Agency," 709.

28. Harvey, *Cosmopolitanism and the Geographies of Freedom*.

29. Henri Lefebvre, *Critique of Everyday Life*, 3 vols. (London: Verso, 2002), 2:36.

30. Ibid., 2:37.

31. Malcolm Miles, *Urban Avant-Gardes: Art, Architecture, and Change* (London: Routledge, 2004), 14.

32. Antonio Gramsci, "Art and the Struggle for a New Civilization," in *The Antonio Gramsci Reader*, ed. David Forgacs (London: Lawrence and Wishart, 1988), 394.

33. Walter Benjamin, "The Author as Producer," *New Left Review* 1, no. 62 (July–August 1970): 83–96.

34. Rendell, *Art and Architecture.*

35. Miles, *Urban Avant-Gardes.*

36. For a useful definition of political ecology that builds on this, see Paul Robbins, *Political Ecology: A Critical Introduction* (Oxford: Blackwell, 2004).

37. See Swyngedouw's transformation of Latour in Erik Swyngedouw, "Modernity and Hybridity: Nature, Regeneracionismo, and the Production of the Spanish Waterscape, 1890–1930," *Annals of the Association of American Geographers* 89, no. 3 (September 1999): 443–65.

38. Neil Smith, *Uneven Development: Nature, Capital, and the Production of Space* (Oxford: Blackwell, 1984).

39. Marx, *Early Writings*, 421.

40. Gramsci's use of "coherent" and "incoherent" will become clearer in the chapter. As Peter Thomas demonstrates in *The Gramscian Moment: Philosopy, Hegemony, and Marxism* (Leiden: Brill, 2009), this differs fundamentally from logical coherence.

41. John Roberts, *Philosophizing the Everyday* (London: Pluto Press, 2006), 69.

1. The Urbanization of Nature

1. Don Mitchell, *The Lie of the Land: Migrant Workers and the California Landscape* (Minneapolis: University of Minnesota Press, 1996).

2. Donna Haraway, *When Species Meet* (Minneapolis: University of Minnesota Press, 2007), 3.

3. For useful summaries, see Noel Castree, *Nature* (Oxford: Blackwell, 2005), and Jamie Lorimer, "Posthumanism/Posthumanistic Geographies," in *International Encyclopedia of Human Geography*, ed. Nigel Thrift and Rob Kitchen (Amsterdam: Elsevier Science, 2009).

4. David Livingstone, *The Geographical Tradition: Episodes in the History of a Contested Enterprise* (Oxford: Blackwell, 1992).

5. Castree, *Nature.*

6. Clarence Glacken, *Traces on the Rhodian Shore: Nature and Culture in Western Thought from Ancient Times to the End of the Eighteenth Century* (Berkeley: University of California Press, 1967); Carl O. Sauer, "The Morphology of Landscape," reprinted in *Human Geography: An Essential Anthology*, ed. John Agnew, David Livingstone, and Alasdair Rogers (Oxford: Blackwell, 1997); Paul Vidal de la Blache, *Principles of Human Geography* (London: Constable, 1918).

7. David Harvey, *Justice, Nature, and the Geography of Difference* (Oxford: Blackwell, 1996), 186–87.

8. Bruno Latour, *Reassembling the Social: An Introduction to Actor-Network Theory* (Oxford: Oxford University Press, 2005); Nigel Thrift, *Non-representational Theory: Space, Politics, Affect* (London: Routledge, 2007); Lorimer, "Posthumanism/Posthumanistic Geographies."

9. Neil Smith, *Uneven Development: Nature, Capital, and the Production of Space* (Oxford: Blackwell, 1984).

10. David Harvey, "Cities or Urbanization?" *City* 1, no. 1 (1996): 50.

11. Ibid.

12. Erik Swyngedouw and Maria Kaika, "The Environment of the City . . . or the Urbanization of Nature," in *A Companion to the City*, ed. Gary Bridge and Sophie Watson (Oxford: Blackwell, 2000), 569.

13. Ibid., 577.

14. Nick Heynen, Maria Kaika, and Erik Swyngedouw, eds., *In the Nature of Cities: Urban Political Ecology and the Politics of Urban Metabolism* (London: Routledge, 2006).

15. Harvey, *Justice, Nature, and the Geography of Difference*, 186.

16. Nik Heynen, Maria Kaika, and Erik Swyngedouw, "Urban Political Ecology: Politicizing the Production of Urban Natures," in *In the Nature of Cities*, ed. Heynen et al., 12.

17. William Cronon, *Nature's Metropolis* (New York: W. W. Norton, 1992); Robert Bullard, *Dumping in Dixie: Race, Class, and Environmental Quality* (Boulder, Colo.: Westview Press, 1990); see also Robert Gottlieb, *Forcing the Spring: The Transformation of the American Environmental Movement* (Washington, D.C.: Island Press, 1993).

18. See Nik Heynen, "Justice of Eating in the City: The Political Ecology of Urban Hunger," and Simon Marvin and Will Medd, "Metabolisms of Obe-City: Flows of Fat through Bodies, Cities, and Sewers," in *In the Nature of Cities*, ed. Heynen et al.

19. Erik Swyngedouw, "Power, Nature, and the City. The Conquest of Water and the Political Ecology of Urbanization in Guayaquil, Ecuador: 1880–1990," *Environment and Planning A* 29 (1997): 311–32; Swyngedouw, "Modernity and Hybridity: Nature, *Regeneracionismo*, and the Production of the Spanish Waterscape, 1890–1930," *Annals of the Association of American Geographers* 89, no. 3 (1999): 443–65; Swyngedouw, *Flows of Power: The Political Ecology of Water and Urbanisation in Ecuador* (Oxford: Oxford University Press, 2004).

20. Maria Kaika, *City of Flows: Modernity, Nature, and the City* (New York: Routledge, 2005); Maria Kaika, "Dams as Symbols of Modernisation: The Urbanisation of Nature between Materiality and Geographical Representation," *Annals of the Association of American Geographers* 96, no. 2 (2006): 276–301.

21. Matthew Gandy, *Concrete and Clay: Reworking Nature in New York City* (Cambridge, Mass.: MIT Press, 2002).

22. Michael Ekers and Alex Loftus, "The Power of Water: Developing Dialogues between Foucault and Gramsci," *Environment and Planning D: Society and Space* 26, no. 4 (2008): 698–718.

23. Alex Loftus, "Working the Socio-Natural Relations of the Urban Waterscape," *International Journal of Urban and Regional Research* 31, no. 1 (2007): 41–59.

24. Smith, *Uneven Development*, 16.

25. Alfred Schmidt, *The Concept of Nature in Marx* (London: New Left Books, 1971).

26. Max Horkheimer and Theodor W. Adorno, *The Dialectic of Enlightenment* (Stanford: Stanford University Press, 2002 [1947]).

27. Martin Jay, *The Dialectical Imagination* (Berkeley: University of California Press, 1973), 256.

28. Quoted in Smith, *Uneven Development*, 22.

29. Smith, *Uneven Development*, 30–31.

30. Ibid., 53.

31. Karl Marx, *Early Writings* (London: Pelican, 1975).

32. As I will demonstrate later, this is crucial to the argument Lukács developed. Moreover, there is real potential for developing a dialogue between the two authors around the concept of nature and its potentials and limitations for a liberatory politics. As Smith claims in *Uneven Development*, "Certain aspects of nature are available to some classes only as a conceptual abstraction, not as a physical partner or opponent in the work process" (42).

33. Noel Castree, "Marxism and the Production of Nature," *Capital and Class* 72 (2000): 5–36; Castree, "Nature," in *The Dictionary of Human Geography*, ed. R. J. Johnston, Derek Gregory, Geraldine Pratt, and Michael Watts, 4th ed. (Oxford: Blackwell, 2000); Castree, *Nature*.

34. Karl Marx, *Capital: Volume 1* (London: Penguin, 1976).

35. Smith, *Uneven Development*, 54.

36. Castree, "Nature," 663 (emphasis added). Importantly, Castree makes this claim in what remains the key reference dictionary for human geography. For countless undergraduates, graduates, and senior academics in geography, this is the definitive statement on the production of nature.

37. Castree, *Nature*, 161.

38. The dualisms that pervade this comment suggest some of the difficulty with reconciling a "non-natural nature" with the "production of nature" thesis. The interpretation is quite simply wrong.

39. Nevertheless, for an example of where Castree presents Smith's arguments with far more care, see Castree, "Marxism and the Production of Nature." Here he writes: "At base, Smith's notion of production is remarkably simple. Utilising a rather orthodox reading of Marx's political economy in which form (value) dominates over content (use-value, concrete labour), he suggests that capitalism does more than merely 'interact with,' 'appropriate' or even 'articulate with' nature" (26). Instead, Smith conveys a much stronger sense of the quantitative and qualitative transformation of nature in capitalist societies.

40. Smith, *Uneven Development*, 55.

41. For a similar argument, see Harvey, *Justice, Nature, and the Geography of Difference*.

42. In an otherwise wonderful account, this appears to be the argument in John Bellamy Foster, *Marx's Ecology: Materialism and Nature* (New York: Monthly Review Press, 2000).

43. Castree, "Marxism and the Production of Nature," 25. This is surely a mischaracterization of at least three of these theorists.

44. Smith, *Uneven Development*, 159, 60.

45. Sarah Whatmore, "Hybrid Geographies: Rethinking the 'Human' in Human Geography," in *Human Geography Today*, ed. Doreen Massey, John Allen, and Philip Sarre (London: Sage, 1999), 25.

46. Lorimer, "Posthumanism/Posthumanistic Geographies."

47. Steve Hinchliffe, *Geographies of Nature: Societies, Environments, Ecologies* (London: Sage, 2007), 51.

48. Noel Castree, "False Antitheses? Marxism, Nature, and Actor-Networks," *Antipode* 34, no. 1 (2002): 111–46.

49. Neil Badmington, "Mapping Posthumanism," *Environment and Planning A* 36 (2004): 1345.

50. Nicholas Gane, "When We Have Never Been Modern, What Is to Be Done? An Interview with Donna Haraway," *Theory, Culture & Society* 23, nos. 7–8 (2006): 135–58.

51. Ibid., 149.

52. Haraway, *When Species Meet*.

53. Neil Smith, "Nature as Accumulation Strategy," *Socialist Register* 43 (2007): 24.

54. Ted Benton, "Marxism and Natural Limits: An Ecological Critique and Reconstruction," *New Left Review* 1, no. 178 (November–December 1989): 51–86; Reiner Grundmann, "The Ecological Challenge to Marxism," *New Left Review* 1, no. 187 (May–June 1991): 103–20; Ted Benton, "Ecology, Socialism, and the Mastery of Nature: A Reply to Reiner Grundmann," *New Left Review* 1, no. 191 (July–August 1992): 55–74.

55. David Harvey, "On the Deep Relevance of a Certain Footnote in Marx's Capital," *Human Geography* 1, no. 2 (2008): 26–31.

56. David Harvey, *Cosmopolitanism and the Geographies of Freedom* (New York: Columbia University Press, 2009), 244–45.

57. Ibid., 245.

58. Ibid., 246–47.

59. Harvey pursues the point further in a revealing exchange with Michael Hardt and Antonio Negri in *Artforum* (November 2009).

2. Sensuous Socio-Natures

1. Preben Kaarsholm, "Moral Panic and Cultural Mobilization: Responses to Transition, Crime, and HIV/AIDS in KwaZulu Natal," *Development and Change* 36, no. 1 (2005): 133–56.

2. Maria Kaika, "Constructing Scarcity and Sensationalising Water Politics: 170 Days That Shook Athens," *Antipode* 35, no. 5 (2003): 919–54. For E. P. Thompson, similarly, writing by candlelight, the interruption of a steady supply of electricity through the general strike served to bring within the household the relations of production rarely considered in daily life: "In our reified mental world, we think we are dependent upon things. What other people do for us is mediated by inanimate objects: the switch,

the water tap, the lavatory chain, the telephone receiver, the cheque through the post." See E. P. Thompson, *Writing by Candlelight* (London: Merlin Press, 1980), 47.

3. This research visit was part of the fieldwork for a DPhil at Oxford University. I was accompanied by Fiona Lumsden and Thulani Ncwane.

4. Karl Marx, *Early Writings* (London: Pelican, 1975), 328.

5. Donna Haraway, *When Species Meet* (Minneapolis: University of Minnesota, 1988), 46.

6. Marx, *Early Writings*, 365.

7. John Bellamy Foster, *Marx's Ecology: Materialism and Nature* (New York: Monthly Review Press, 2000).

8. Haraway, *When Species Meet*, 1.

9. For important exceptions, see, among others, Paul Burkett, *Marx and Nature: A Red and Green Perspective* (New York: St. Martin's Press, 1999); Foster, *Marx's Ecology*; David Harvey, *Cosmopolitanism and the Geographies of Freedom* (New York: Columbia University Press, 2009).

10. Ted Benton, "Marxism and Natural Limits: An Ecological Critique and Reconstruction," *New Left Review* 1, no. 178 (November–December 1989).

11. John Holloway, *Change the World without Taking Power* (London: Pluto Press, 2002).

12. David Harvey, "Population, Resources, and the Ideology of Science," *Economic Geography* 50 (1974): 256–77; Foster, *Marx's Ecology*.

13. Harvey, *Cosmopolitanism and the Geographies of Freedom*.

14. Marx, *Early Writings*, 350.

15. For one of the clearest and most conclusive rebuttals of such a position, see Foster, *Marx's Ecology*.

16. Whereas Foster goes to great lengths to uncover the true Baconian roots of such an expression and in the process demonstrates how this "domination" should not be read in a mechanist sense, Lefebvre's reading often lacks such nuance.

17. Foster, *Marx's Ecology*; Burkett, *Marx and Nature*.

18. Foster, *Marx's Ecology*, 231.

19. Andrew Feenberg, *Lukács, Marx, and the Sources of Critical Theory* (Oxford: Oxford University Press, 1981), 221.

20. Georg Lukács, *History and Class Consciousness* (London: Merlin Press, 1971), 134.

21. Karl Marx and Friedrich Engels, *The German Ideology* (London: Lawrence and Wishart, 1970), 37.

22. Marx, *Early Writings*, 381.

23. Ludwig Feuerbach, *Principles of the Philosophy of the Future* (Indianapolis: Bobbs-Merrill, 1966).

24. Kate Soper, *Humanism and Anti-humanism* (London: Hutchinson, 1986), 31.

25. It is for this reason that Engels claims that "the German working class movement is the inheritor of German classical philosophy." This claim has been a touchstone

for debates over the meaning of a marxist philosophy. In quite different ways, it is central to the theoretical claims of both Althusser and Gramsci (see Peter D. Thomas, *The Gramscian Moment: Philosophy, Hegemony, and Marxism* [Leiden: Brill, 2009]).

26. Foster, *Marx's Ecology*, 72–77; István Mészáros, *Marx's Theory of Alienation* (London: Merlin Press, 2006).

27. Richard Levins and Richard Lewontin, *The Dialectical Biologist* (Cambridge, Mass.: Harvard University Press, 1985), 105–6.

28. For a good discussion, see David Harvey, *Limits to Capital*, 2nd ed. (London: Verso, 1999), 14–16.

29. Neil Smith, *Uneven Development: Nature, Capital, and the Production of Space* (Oxford: Blackwell, 1984), 55.

30. Marx, *Early Writings*, 324.

31. Smith, *Uneven Development*.

32. Foster, *Marx's Ecology*.

33. Although work on urban agriculture (see, for example, Jason W. Moore, "Why Farm the City? Theorizing Urban Agriculture through a Lens of Metabolic Rift," *Cambridge Journal of Regions, Economy and Society* 3 [2010]: 191–207), provides suggestive possibilities.

34. David Harvey, *Justice, Nature, and the Geography of Difference* (Oxford: Blackwell, 1996), 185–86.

35. Marx, *Early Writings*, 328–29.

36. For a discussion, see Foster, *Marx's Ecology*.

37. Karl Marx, *Grundrisse* (London: Pelican, 1973), 489.

38. Smith, *Uneven Development*, 30–31.

39. David McLellan, *Karl Marx: His Life and Works* (London: Macmillan, 1973), 114.

40. I have already noted this in reference to Lefebvre's detection of an ethical and an aesthetic critique of everyday life within Marx.

41. Marx, *Early Writings*, 353.

42. Feenberg, *Lukács, Marx, and the Sources of Critical Theory*, 218.

43. Indeed, a form of this practical materialism could later be seen in somewhat different ways in the work of the Annales School or in the cultural geography of Carl Sauer.

44. Lukács, *History and Class Consciousness*, 137.

45. Ibid., 140.

46. Henri Lefebvre, *Everyday Life in the Modern World* (London: Continuum, 2002), 204.

47. McLellan, *Karl Marx*, detects a debt to Schiller in the importance Marx affords to the model of creative practice. He goes on to note how at the time of writing the Paris Manuscripts as well as discovering new forms of associational practice through his interaction with communist artisans in Paris, Marx's closest friends at the time were poets (151). These newfound friends shaped his particular sense of communism

and also his engagement with nature. The privileging of the senses, in turn within this philosophy of praxis, is related to the influence of both Epicurus and Feuerbach.

48. Feenberg, *Lukács, Marx, and the Sources of Critical Theory*, 218.

49. Reiner Grundmann, "The Ecological Challenge to Marxism," *New Left Review* 1, no. 187 (May–June 1991): 111.

50. Harvey, *Justice, Nature, and the Geography of Difference*, 189.

51. Ted Benton, "Ecology, Socialism, and the Mastery of Nature: A Reply to Reiner Grundmann," *New Left Review* 1, no. 191 (July–August 1992): 64.

52. Feenberg, *Lukács, Marx, and the Sources of Critical Theory*, 217.

53. Here Foster's excavation of the Baconian roots of the concept shows the shallowness to some interpretations. See Foster, *Marx's Ecology*.

54. Feenberg, *Lukács, Marx, and the Sources of Critical Theory*, 217–18.

55. Ibid., 218.

56. Smith, *Uneven Development*.

57. Feenberg, *Lukács, Marx, and the Sources of Critical Theory*, 217.

58. Ibid., 219.

59. On this connection, see Foster, *Marx's Ecology*.

60. All references to the eleven theses are taken from Marx, *Early Writings*, 421–23.

61. Marx and Engels, *The German Ideology*, 63.

62. In this regard, they are at the heart of Gramsci's integral humanism, which, as Thomas shows, bears no relation to the straw humanism criticized by Althusser. See Thomas, *The Gramscian Moment*.

63. See, for example, Steve Hinchliffe, *Geographies of Nature: Societies, Environments, Ecologies* (London: Sage, 2007), on how nature is performed or enacted through sensuous experience.

64. Jane Rendell, *Art and Architecture: A Place Between* (London: I. B. Tauris, 2006), 11.

65. Malcolm Miles, *Urban Avant-Gardes: Art, Architecture, and Change* (London: Routledge, 2004).

66. For more details, see http://www.softhook.com/ (accessed December 21, 2009).

67. Erik Swyngedouw and Maria Kaika, "The Environment of the City . . . or the Urbanization of Nature," in *A Companion to the City*, ed. Gary Bridge and Sophie Watson (Oxford: Blackwell, 2000).

68. For more details, see http://socialtapestries.net/feralrobots/ (accessed December 21, 2009).

69. See http://www.bruno-latour.fr/virtual/index.html.

3. Cyborg Consciousness

1. Michael Löwy, *Georg Lukács: From Romanticism to Bolshevism* (London: New Left Books, 1979).

2. Slavoj Žižek, "Postface: Georg Lukács as the Philosopher of Leninism," in Georg

Lukács, *A Defence of History and Class Consciousness: Tailism and the Dialectic* (London: Verso, 2000), 178.

3. In English, as Georg Lukács, *A Defence of History and Class Consciousness: Tailism and the Dialectic*.

4. Paul Burkett, "Lukács on Science: A New Act in the Tragedy," *Economic and Political Weekly* 36, no. 48 (December 1–7, 2001): 4485–89.

5. Löwy, *Georg Lukács*, 170.

6. Lukács, *History and Class Consciousness*, 89.

7. Neil Smith, "Antinomies of Space and Nature in Henri Lefebvre," in *Philosophy and Geography 2: The Production of Public Space*, ed. Andrew Light and Jonathan M. Smith (Lanham, Md.: Rowman and Littlefield, 1998), 55.

8. Richard Peet, *Modern Geographical Thought* (Oxford: Blackwell, 1998), 91.

9. Gearóid Ó Tuathail, *Critical Geopolitics* (London: Routledge, 1996), 147.

10. Marshall Berman, *Adventures in Marxism* (London: Verso, 1999), 205.

11. Löwy, *Georg Lukács*, 65.

12. Berman, *Adventures in Marxism*.

13. Georg Lukács, *Theory of the Novel* (London: Merlin Press, 1971), 29.

14. Ibid., 152.

15. Löwy, *Georg Lukács*.

16. Ibid., 123.

17. Georg Lukács, *Tactics and Ethics* (New York: Harper Torchbooks, 1972), 5.

18. Lukács, *History and Class Consciousness*, 1.

19. Löwy, *Georg Lukács*.

20. Lukács, *Tactics and Ethics*, 27.

21. Nevertheless, the presentation of the chapters is not strictly a chronological one.

22. Žižek, "Postface."

23. Terry Eagleton, *Ideology: An Introduction* (London: Verso, 1991).

24. Lukács, *History and Class Consciousness*, 88.

25. For Marx: "The propertied class and the class of the proletariat present the same human self-alienation. But the former class finds in this self-alienation its confirmation and its good, its own power: it has in it a semblance of human existence. The class of the proletariat feels annihilated in its self-alienation; it sees in it its own powerlessness and the reality of an inhuman existence" (David McLellan, *Karl Marx: His Life and Works* [London: Macmillan, 1973], 134).

26. Ibid., 86.

27. Ibid., xliii.

28. Martin Jay's intellectual history, *Marxism and Totality*, perhaps not surprisingly, begins with Lukács as the seminal marxist writer on totality; indeed, it is the point of view of totality, so well captured in the work of Rosa Luxemburg, that Lukács finds to be the decisive difference between marxism and bourgeois thought (27).

29. Lukács, *History and Class Consciousness*, 3.

30. Ibid., 9.
31. Ibid., 164.
32. Ibid., 260.
33. Karl Marx, *Early Writings* (London: Pelican, 1975), 421.
34. Löwy, *Georg Lukács*, too, captures this, arguing that Althusser's rejection of *History and Class Consciousness* as left-wing humanism misses the revolutionary humanism in the work (182). As with Marx, this relies on: the demystification of reified forms; a critique of the inhuman effects of capitalism; and human emancipation through socialist revolution.
35. Ibid.
36. Ibid., 69.
37. Ibid., 165.
38. Ibid., 174. This also echoes Zola's description of Lantier's arrival in Montsou in the novel *Germinal*. On questioning who owns the mines in the town, his voice takes "on a kind of religious awe; it was as if he had spoken of some untouchable tabernacle which concealed the crouching greedy god to whom they all offered up their flesh, but whom they had never seen."
39. Ibid., 197.
40. Nancy C. M. Hartsock, *Money, Sex, and Power: Toward a Feminist Historical Materialism* (London: Longman, 1983), 118.
41. Hartsock, in *Money, Sex, and Power*, refers to the "sexual" division of labor as opposed to the gendered division of labor as part of her refusal to separate "nature and nurture" and to emphasize that this is more than a social division of labor (233). This is not altogether unproblematic and is rejected by most other feminist standpoint theorists.
42. Ibid.
43. Caroline New, "Man Bad, Woman Good? Essentialisms and Ecofeminisms," *New Left Review* 1, no. 216 (1997): 79–93. It should be noted that New's elision of ecofeminism and standpoint theory collapses significant differences and is unhelpful.
44. Donna J. Haraway, *Modest_Witness@Second_Millennium.FemaleMan©_Meets _Oncomouse™* (New York: Routledge, 1996), 198–99.
45. Haraway would quite possibly refuse such a label, although it has been a crucial part of some of her writings. I do not wish to collapse the differences among any of these authors.
46. Chela Sandoval, "New Sciences: Cyborg Feminism and the Methodology of the Oppressed," in *The Cyborg Handbook*, ed. Chris Hables Gray (New York: Routledge, 1995), 375.
47. Kevin Floyd, *The Reification of Desire: Towards a Queer Marxism* (Minneapolis: University of Minnesota Press, 2009). I am grateful to Mike Ekers for making this connection.
48. Henri Lefebvre, *The Sociology of Marx* (Harmondsworth: Penguin, 1972), 36.
49. John Roberts, *Philosophizing the Everyday* (London: Pluto Press, 2006).

50. Lukács, *History and Class Consciousness*, 51.
51. Ibid., 79.
52. Roberts, *Philosophizing the Everyday*.
53. Lukács, *History and Class Consciousness*, 149 (emphasis added).
54. Ibid., 162.
55. Löwy, *Georg Lukács*, 191.
56. Roberts, *Philosophizing the Everyday*, 35.
57. Fredric Jameson, "History and Class Consciousness as an 'Unfinished Project,'" in *The Feminist Standpoint Theory Reader*, ed. Sandra Harding (London: Routledge, 2004), 144–45.
58. Gareth Stedman Jones, "The Marxism of the Early Lukács," in *Western Marxism: A Critical Reader*, ed. Gareth Stedman Jones (London: New Left Books, 1978), 44.
59. Lukács, *Tailism and the Dialectic*, 59. For Lukács, the moment is "a situation whose duration may be longer or shorter, but which is distinguished from the process that leads up to it in that it forces together the essential tendencies of that process, and demands that a *decision* be taken over the future direction of the process" (55). In part, as Žižek shows, we find a far more Gramscian Lukács here, one who emphasizes the conjunctural and the contingent. Proletarian class consciousness requires not simply a dialectical understanding of the totality and of its own subject–object status within this (an understanding of the organic for Gramsci), but also a careful survey of the terrain ahead and a confidence to assert, against the odds, that the time for revolution is right. It is in this regard, Žižek argues, that we should understand Lukács's theory of the Party. And it is in this regard that Lukács might assume the mantle given to him as the ultimate philosopher of Leninism. Later I will argue that, as the Situationists and Lefebvre were to observe, the generation of such moments within the urban fabric can force revolutionary ruptures in the contemporary instant. Indeed, these provide far more realistic vehicles than the Leninist party. If the reified world constitutes the immediacy to be transcended by proletarian class consciousness, this is not to deny the importance of this everyday experience for Lukács's overall outlook.
60. Ibid., 54.
61. Ibid., 66, 41.
62. Andrew Feenberg, *Lukács, Marx and the Sources of Critical Theory* (Oxford: Oxford University Press, 1981), 154.
63. Interview with Löwy in Eva L. Corredor, *Lukács after Communism: Interviews with Contemporary Intellectuals* (Durham: Duke University Press, 1997), 27.
64. Furthermore, Löwy refutes Stedman Jones's claim, also represented in the quotation from Roberts, that a pure consciousness is crudely juxtaposed with an impure, actual consciousness, by emphasizing the clear gradations within the Lukácsean understanding of class consciousness.
65. Jameson, "History and Class Consciousness as an 'Unfinished Project,'" 145–46.
66. Lukács, *Tailism and the Dialectic*, 56.

67. See Susan Hekman, "Truth and Method: Feminist Standpoint Theory Revisited," in *The Feminist Standpoint Theory Reader*, ed. Sandra Harding (New York: Routledge, 2004), for some of these criticisms that she puts down, in large part, to the adoption of a "discredited" marxist methodology).

68. Žižek, "Postface," 168.

69. Ibid., 170.

70. Lukács, *Tactics and Ethics*, 142.

71. Antonio Gramsci, *Selections from the Prison Notebooks* (London: Lawrence and Wishart, 1971), 428.

72. Ibid., 426.

73. Lukács, *Tactics and Ethics*, 139.

74. Ibid., 136.

75. See Alex Loftus, "Reification and the Dictatorship of the Water Meter," *Antipode* 38, no. 5 (2006): 1023–44.

76. Robert M. Young, "Science *is* Social Relations," *Radical Science Journal* 5 (1977); Young, *Darwin's Metaphor: Nature's Place in Victorian Culture* (Cambridge: Cambridge University Press, 1985).

77. Stedman Jones, "The Marxism of the Early Lukács," 44.

78. Ibid., 45.

79. Lukács, *History and Class Consciounsess*, 10.

80. Alfred Schmidt, *The Concept of Nature in Marx* (London: New Left Books, 1971).

81. Ibid., 24.

82. Ibid., 132.

83. Andrew Arato and Paul Breines, *The Young Lukács and the Origins of Western Marxism* (London: Pluto Press, 1979), 180.

84. Feenberg, *Lukács, Marx, and the Sources of Critical Theory*; Jay, *Marxism and Totality*, 116.

85. Feenberg, *Lukács, Marx, and the Sources of Critical Theory*, 211.

86. Gramsci, *Selections from the Prison Notebooks*, 448. This is a C-text. In the A-text (appearing earlier in the notebooks Gramsci; see Gramsci Notebook 4, §43. Antonio Gramsci, *Prison Notebooks*, 3 vols. [New York: Columbia University Press, 1996], 2:192–93), claims to know Lukács's theories only "very vaguely."

87. Steven Vogel, *Against Nature: The Concept of Nature in Critical Theory* (Albany: State University of New York Press, 1997).

88. Ibid., 31.

89. Burkett, "Lukács on Science," 4485.

90. Ibid., 4487.

91. Young, "Science *is* Social Relations."

92. This refers to Burkett's judgment of *Tailism and the Dialectic* in "Lukács on Science."

93. John Rees, introduction to *Tailism and the Dialectic*, 1–2.

94. Lukács, *Tailism and the Dialectic*, 102.

95. Ibid., 106.

96. Vogel, *Against Nature*.

97. Lukács, *History and Class Consciousness*, 234.

98. Ibid., xvii (emphasis added).

99. See Lucio Colletti, *Hegel and Marx* (London: New Left Books, 1973); Schmidt, *The Concept of Nature in Marx*.

100. Donna Haraway, *The Donna Haraway Reader* (London: Routledge, 2004), 139.

101. Erik Swyngedouw, "Circulations and Metabolisms: (Hybrid) Natures and (Cyborg) Cities," in *Technonatures: Environments, Technologies, Spaces, and Places in the Twenty-first Century*, ed. Damian White and Chris Wilbert (Waterloo, Ont.: Wilfrid Laurier University Press, 2009).

102. White and Wilbert, eds., *Technonatures*.

103. Ibid., 100.

104. David Harvey, *Justice, Nature, and the Geography of Difference* (Oxford: Blackwell, 1996).

105. Lukács, *History and Class Consciousness*, 165.

106. Ibid., 164.

107. Maria Kaika and Erik Swyngedouw, "Fetishising the Modern City: The Phantasmagoria of Urban Technological Networks," *International Journal of Urban and Regional Research* 24, no. 1 (2000): 120–38.

108. Maria Kaika, "Constructing Scarcity and Sensationalising Water Politics: 170 Days That Shook Athens," *Antipode* 35, no. 5 (2003): 919–54.

109. Lukács, *History and Class Consciousness*, 165.

110. Ibid., 168.

111. Ibid., 169.

112. Claude Meillassoux, *Maidens, Meal, and Money: Capitalism and the Domestic Community* (Cambridge: Cambridge University Press, 1983).

113. Harold Wolpe, "Capitalism and Cheap Labour Power: From Segregation to Apartheid," *Economy and Society* 1, no. 4 (1972): 425–56.

114. Erik Swyngedouw, "Privatising H2O: Turning Local Waters into Global Money," *Austrian Journal of Development Studies* 19, no. 4 (2003): 10–33.

115. Žižek, "Postface," 170.

116. Haraway, *Modest_Witness@Second_Millennium.FemaleMan©_Meets_Oncomouse™*, 198–99.

117. Damian White and Chris Wilbert, "Technonatural Time/Spaces," *Science as Culture* 15, no. 2 (2006): 100.

118. While the possibility is recognized in Mike Michael's *Technoscience and Everyday Life* through the contrasting examples of Lefebvre and de Certeau, this is quickly passed off as a form of utopianism.

119. Marianne de Laet and Annemarie Mol, "The Zimbabwe Bush Pump: Mechanics of a Fluid Technology," *Social Studies of Science* 30, no. 2 (2000): 225–63.

120. Scott Kirsch and Don Mitchell, "The Nature of Things: Dead Labour, Nonhuman Actors, and the Persistence of Marxism," *Antipode* 36 (2004): 687–706.

4. When Theory Becomes a Material Force

1. Peter Thomas's utterly outstanding *The Gramscian Moment: Philosophy, Hegemony, and Marxism* (Leiden: Brill, 2009), quoting Gramsci, frames the notebooks around an absolute historicism, absolute humanism, and absolute immanence. Again, I think this can be read as a historically and geographically situated immanent critique.

2. John Holloway, *Change the World without Taking Power* (London: Pluto Press, 2002), 156.

3. Antonio Gramsci, *Selections from the Prison Notebooks* (London: Lawrence and Wishart, 1971), 9.

4. Ibid., 323.

5. Neil Smith, *Uneven Development: Nature, Capital, and the Production of Space* (Oxford: Blackwell, 1984).

6. Gramsci, *Selections from the Prison Notebooks*, 344.

7. Michael Hardt and Antonio Negri, *Empire* (Cambridge, Mass.: Harvard University Press, 2000), 58.

8. Bertolt Brecht, *Mother Courage and Her Children* (London: Methuen, 1962), 41.

9. Richard Day, *Gramsci Is Dead: Anarchist Currents in the Newest Social Movements* (London: Pluto Press, 2005).

10. Gramsci, *Selections from the Prison Notebooks*, 462.

11. Stuart Hall, "The Problem of Ideology: Marxism without Guarantees," *Journal of Communication Studies* 10, no. 2 (1986): 28–44.

12. Quintin Hoare and Geoffrey Nowell-Smith, Introduction to Gramsci, *Selections from the Prison Notebooks*, xii.

13. Maurice Finocchiaro, *Gramsci and the History of Dialectical Thought* (Cambridge: Cambridge University Press, 1988), 89.

14. Ibid., 93.

15. Wolfgang Fritz Haug, "Gramsci's 'Philosophy of Praxis,'" *Socialism and Democracy* 14, no. 1 (Summer 2000): 19.

16. Ibid., 2.

17. Gramsci, *Selections from the Prison Notebooks*, 381. The phrase is actually from Engels's *Ludwig Feuerbach and the End of Classical German Idealism*.

18. Ibid., 367.

19. Haug, "Gramsci's 'Philosophy of Praxis.'"

20. Stuart Hall, "Gramsci's Relevance for the Study of Race and Ethnicity," *Journal of Communication Studies* 10, no. 2 (1986): 20.

21. Gramsci, *Selections from the Prison Notebooks*, 424.

22. See Kate Crehan, *Gramsci: Culture and Anthropology* (London: Pluto Press, 2003).

23. Gramsci, *Selections from the Prison Notebooks*, 326.

24. Ibid., 233.
25. Ibid., 333.
26. Ibid.
27. Ibid., 330.
28. Ibid., 418.
29. Ibid., 335.
30. Andrew Feenberg, *Lukács, Marx, and the Sources of Critical Theory* (Oxford: Oxford University Press, 1981), 152.
31. Crehan, *Gramsci: Culture and Anthropology.*
32. Haug, "Gramsci's 'Philosophy of Praxis.'"
33. Peter Thomas, "Gramsci and the Political: From the State as 'Metaphysical Event' to Hegemony as 'Philosophical Fact,'" *Radical Philosophy* 153 (January–February 2009): 27–37.
34. Gramsci, *Selections from the Prison Notebooks,* 352.
35. Ibid.
36. Karl Marx, *Early Writings,* (London: Pelican), 422.
37. Gramsci, *Selections from the Prison Notebooks,* 350.
38. Ibid., 435.
39. Ibid., 448.
40. Ibid., 407.
41. Benedetto Fontana, "The Concept of Nature in Gramsci," *Philosophical Forum* 27, no. 3 (1996): 221.
42. Ibid., 230.
43. Gramsci, *Selections from the Prison Notebooks,* 446.
44. Ibid., 358.
45. Heather Hughes, "The City Closes In: The Incorporation of Inanda into Metropolitan Durban," in *The People's City: African Life in Twentieth-Century Durban,* ed. Paul Maylam and Ian Edwards (Pietermaritzburg: University of Natal Press, 1996), 299–306.
46. Heather Hughes, "Violence in Inanda, August 1985," *Journal of Southern African Studies* 13 (1987): 347–68.
47. David Hemson, "'For sure you are going to die!' Political Participation and the Comrade Movement in Inanda, KwaZulu Natal," *Social Dynamics* 22 (1996): 74–105.
48. Ibid., 78.
49. Heather Hughes, introduction to *Imijondolo: A Photographic Essay of Forced Removals in South Africa,* photographs by Omar Badsha; foreword by Bishop Desmond Tutu (Johannesburg: Afrapic, 1985), i–vi.
50. Hughes, "Violence in Inanda, August 1985"; Fatima Meer, *Unrest in Natal, August 1985* (Durban: Institute for Black Research, 1985).
51. Surplus People Project Reports, *Forced Removals in South Africa,* vol. 4 (Cape Town: Natal SPP, 1983).

52. Hughes, "The City Closes In."

53. Paul Maylam, "The Struggle for Space in Twentieth-Century Durban," in *The People's City: African Life in Twentieth-Century Durban*, ed. Paul Maylam and Ian Edwards, 1–32.

54. Hughes, "The City Closes In."

55. David Harvey, *The Urban Experience* (Oxford: Basil Blackwell, 1989).

56. Erik Swyngedouw and Maria Kaika, "The Environment of the City . . . or the Urbanization of Nature," in *A Companion to the City*, ed. Gary Bridge and Sophie Watson (Oxford: Blackwell, 2000), 569; Maria Kaika, *City of Flows: Modernity, Nature, and the City* (New York: Routledge, 2005).

57. Meer, *Unrest in Natal, August 1985*.

58. Doug Hindson, "Inanda: Report on the State of the Environment and Development of the Durban Metropolitan Area," City of Durban, North Central Council, Physical Environment Planning Unit, 1996. This was confirmed in a community workshop held in Bhambayi in 2003.

59. Garth Seneque, "The Inanda Scheme: A Briefing," *Work in Progress* 20 (1981): 8–15.

60. Dan Smit, "The Urban Foundation: Transformation Possibilities," *Transformation* 18 (1992): 35–42.

61. Patrick Bond, *Commanding Heights and Community Control: New Economics for a New South Africa* (Johannesburg: Ravan Press, 1991).

62. Smit, "The Urban Foundation."

63. Seneque, "The Inanda Scheme."

64. Hindson, "Inanda."

65. David Hemson stressed this point in several conversations I shared with him.

66. Hughes, "The City Closes In"; "Violence in Inanda, August 1985"; introduction.

67. Ari Sitas, "The Making of the Comrades Movement in Natal, 1985–1991," *Journal of Southern African Studies* 18, no. 3 (1992): 629–41.

68. Meer, *Unrest in Natal, August 1985*.

69. Hughes, "Violence in Inanda, August 1985."

70. Ibid., 347.

71. Gillian Hart, *Disabling Globalization: Places of Power in Post Apartheid South Africa* (Pietermaritzburg: University of Natal Press, 2002).

72. Jeremy Seekings, *The UDF: A History of the United Democratic Front in South Africa, 1983–1991* (Oxford: James Currey, 2000).

73. David Harvey, "On the Deep Relevance of a Certain Footnote in Marx's Capital," *Human Geography* 1, no. 2 (2008): 26–31.

74. Sitas, "The Making of the Comrades Movement in Natal."

75. Hemson, "'For sure you are going to die!'"

76. Gramsci, *Selections from the Prison Notebooks*, 5–23.

77. Preben Kaarsholm, "Moral Panic and Cultural Mobilization: Responses to

Transition, Crime, and HIV/AIDS in KwaZulu Natal," *Development and Change* 36 (2005): 133–56.

78. Thulani Ncwane, "Fighting for All Our People to Stay Alive," *De Zuid Afrikaan*, 1990.

79. Hemson, "'For sure you are going to die!'"

80. Ibid.

81. Interview with Thulani Ncwane, March 3, 2003.

82. Interview with Thulani Ncwane, November 10, 2002.

83. Hemson, "'For sure you are going to die!'"

84. Kaarsholm, "Moral Panic and Cultural Mobilization."

85. Gramsci, *Selections from the Prison Notebooks*, 238.

86. See Alex Loftus, "Reification and the Dictatorship of the Water Meter," *Antipode* 38, no. 5 (2006): 1023–44.

87. Business Partners for Development, *Cost Recovery in Partnership: Results, Attitudes, Lessons, and Strategies*, http://www.bpd-waterandsanitation.org/english/docs/costrec.pdf.

88. For a deeper discussion, see David Hemson, "Built to Fly? Or Set Up to Fail? A Review of the Business Partners for Development Pilot Projects in KwaZulu Natal," paper prepared for the Water Research Council, Pretoria, 2003.

89. Interview with Reg Bailey, head of research and development at eThekwini Water Services, February 14, 2003; interview with Neil Macleod, head of water and sanitation, eThekwini Water Services, March 26, 2003.

90. Interview with Manu Pillay, water quality and education manager, December 11, 2002; interview with Reg Bailey, February 14, 2003.

91. David McDonald and John Pape, *Cost Recovery and the Crisis of Service Delivery in South Africa* (Pretoria: HSRC, 2002).

92. Ashwin Desai, *We Are the Poors* (New York: Monthly Review Press, 2002).

93. Jaap de Visser, "From the Courts: Manqele v. Durban Transitional Metropolitan Council, Disconnection of Water Supplies," *Local Government Law Bulletin* 1 (2001).

94. Interview with Neil Macleod, March 26, 2003; interview with councillor S. Khuzwayo, head of infrastructure subcommittee, eThekwini Municipality, April 11, 2003.

95. Interview with Neil Macleod, March 26, 2003.

96. Interview with Karine LeMaux, representative from Vivendi Water in Durban, April 15, 2003.

97. Business Partners for Development, *Cost Recovery in Partnership*.

98. Interview with former activist Thulani Ndlovu, January 30, 2003.

99. African National Congress, *Local Government Election Manifesto: 2000*. http://www.anc.org.za/docs/manifesto/2000/ANC LOCAL GOVERNMENT ELECTI . . . pdf (accessed September 5, 2011).

100. Gramsci, *Selections from the Prison Notebooks*, 310.

101. John Bellamy Foster, *Marx's Ecology: Materialism and Nature* (New York: Monthly Review Press, 2000), 246.

5. Cultural Praxis as the Production of Nature

1. Henri Lefebvre, *Critique of Everyday Life*, 3 vols. (London: Verso, 2002), 2:73.

2. John Roberts, *Philosophizing the Everyday* (London: Pluto Press, 2006), 67.

3. Neil Smith, *Uneven Development: Nature, Capital and the Production of Space* (Oxford: Blackwell, 1984); Damian White and Chris Wilbert, eds., *Technonatures: Environments, Technologies, Spaces, and Places in the Twenty-first Century* (Waterloo, Ont.: Wilfrid Laurier University Press, 2009).

4. Ed Soja, *Postmodern Geographies* (London: Verso, 1989). Henri Lefebvre, *The Production of Space* (Oxford: Blackwell, 1991).

5. For brief examples, see David Harvey, *Social Justice and the City* (Baltimore: Johns Hopkins University Press, 1973); Smith, *Uneven Development*; and Soja, *Postmodern Geographies*. For key works that reflect this change, see Ed Soja, *Thirdspace* (Oxford: Blackwell, 1996); and Rob Shields, *Lefebvre, Love, and Struggle* (New York: Routledge, 1999).

6. Stefan Kipfer, Kanishka Goonewardena, Christian Schmid, and Richard Milgrom, "On the Production of Henri Lefebvre," in *Space, Difference, Everyday Life: Reading Henri Lefebvre*, ed. Goonewardena, Kipfer, Milgrom, and Schmid (New York: Routledge, 2008).

7. Andrew Shmuely, "Totality, Hegemony, Difference: Henri Lefebvre and Raymond Williams," in *Space, Difference, Everyday Life: Reading Henri Lefebvre*, ed. Goonewardena, Kipfer, Milgrom, and Schmid.

8. Kipfer et al., "On the Production of Henri Lefebvre," 3.

9. Stuart Elden, *Understanding Henri Lefebvre: Theory and the Possible* (London: Continuum, 2004), 6.

10. For a counterpoint, see Geoffrey Waite, "Lefebvre without Heidegger: Left-Heideggerianism *qua contradictio in adiecto*," in *Space, Difference, Everyday Life: Reading Henri Lefebvre*, ed. Goonewardena, Kipfer, Milgrom, and Schmid.

11. Andy Merrifield, "The Whole and the Rest: Remi Hess and *les lefebvriens francais*," *Environment and Planning D: Society and Space* 27 no. 5 (2009): 936–49.

12. Malcolm Miles, *Urban Avant-Gardes: Art, Architecture, and Change* (London: Routledge, 2004); Jane Rendell, *Art and Architecture: A Place Between* (London: I. B. Tauris, 2006).

13. Roberts, *Philosophizing the Everyday*.

14. For Waite, in "Lefebvre without Heidegger," this is precisely where Lefebvre's legacy should remain.

15. David Harvey, *Limits to Capital* (London: Verso, 1999); Smith, *Uneven Development*.

16. Kipfer et al., "Producing Henri Lefebvre."

17. Harvey, *Limits to Capital*, 447. Although more recently, he has sought a more nuanced integration of Lefebvrian insights. See David Harvey, *Cosmopolitanism and the Geographies of Freedom* (New York: Columbia University Press, 2009).

18. Stefan Kipfer, "How Lefebvre Urbanized Gramsci," in *Space, Difference, Everyday Life: Reading Henri Lefebvre*, ed. Goonewardena, Kipfer, Milgrom, and Schmid.

19. Alice Kaplan and Kristin Ross, "Everyday Life," special issue of *Yale French Studies*, no. 75 (1987).

20. Again, the *Theses on Feuerbach* can be seen as crucial to Lefebvre's work and his guiding vision: "Philosophy makes itself world: it makes the world and the world is made through it. The world is produced to the exact measure whereby philosophy is realized, and realizing, becomes world. Philosophers have interpreted the world: now it must be changed; can this change be accomplished without philosophy," in *Henri Lefebvre: Key Writings*, ed. Stuart Elden, Eleonore Kofman, and Elizabeth Lebas (London: Continuum, 2003).

21. Elden, *Understanding Henri Lefebvre*.

22. Lefebvre, *Critique of Everyday Life*, 1:179.

23. For excerpts, see *Henri Lefebvre: Key Writings*, ed. Elden, Kofman, and Lebas.

24. Lefebvre, *Critique of Everyday Life*, vol. 2; Andy Merrifield, *Henri Lefebvre: A Critical Introduction* (New York: Routledge, 2006).

25. Quoted in Elden, *Understanding Henri Lefebvre*.

26. Soja, *Thirdspace*.

27. Shields, *Lefebvre, Love, and Struggle*.

28. Elden, *Understanding Henri Lefebvre*, 37.

29. Christian Schmid, "Henri Lefebvre's Theory of the Production of Space: Towards a Three-Dimensional Dialectic," in *Space, Difference, Everyday Life: Reading Henri Lefebvre*, ed. Goonewardena, Kipfer, Milgrom, and Schmid.

30. Lefebvre, *The Production of Space*, 406.

31. Lefebvre, *Key Writings*, 16.

32. Schmid, "Henri Lefebvre's Theory of the Production of Space," 32.

33. Lefebvre, *Key Writings*, 50.

34. Lefebvre, *Critique of Everyday Life*, 2:98–99.

35. Kipfer, "How Lefebvre Urbanized Gramsci," 198.

36. Lefebvre, *Everyday Life in the Modern World*, 12.

37. Lefebvre, *Critique of Everyday Life*, 2:23.

38. Ibid., 3:125.

39. Ibid., 1:124.

40. Ibid., 2:87.

41. Kipfer, "How Lefebvre Urbanized Gramsci."

42. Lefebvre, *Critique of Everyday Life*, 2:21 and 87.

43. Lefebvre, *Key Writings*, 31.

44. Lefebvre, *Critique of Everyday Life*, 2:68.

45. Kaplan and Ross, "Everyday Life."

46. Lefebvre, *Critique of Everyday Life*, 2:16.

47. Roberts, *Philosophizing the Everyday*. See Kanishka Goonewardena, "Marxism and Everyday Life: On Henri Lefebvre, Guy Debord, and Some Others," in *Space, Difference, Everyday Life: Reading Henri Lefebvre*, ed. Goonewardena, Kipfer, Milgrom, and Schmid, for a further rich contextualization of Lefebvre's critique.

48. Roberts, *Philosophizing the Everyday*, 38.

49. Lefebvre writes: "Let us simply say about daily life that it has always existed but permeated with values, with myths. The word *everyday* designates the entry of this daily life into modernity: the everyday as an object of programming (*d'une programmation*), whose unfolding is imposed by the market, by the system of equivalences, by marketing and advertisements. As to the concept of 'everydayness,' it stresses the homogenous, the repetitive, the fragmentary in everyday life" (*Le Monde*, Sunday, December 19, 1982, pp. ix, x, quoted in Henri Lefebvre, "Toward a Leftist Cultural Politics: Remarks Occasioned by the Centenary of Marx's Death," in *Marxism and the Interpretation of Culture*, ed. Cary Nelson and Lawrence Grossberg (Urbana: University of Illinois Press, 1988), 87; see also Roberts, *Philosophizing the Everyday*; Goonewardena, "Marxism in Everyday Life."

50. Roberts, *Philosophizing the Everyday*, 67.

51. Ibid., 42.

52. Lefebvre, *Critique of Everyday Life*, 2:348.

53. Ibid., 343.

54. Ibid., 356.

55. Merrifield, *Henri Lefebvre*, 28.

56. Ibid., 17, 18.

57. Lefebvre, *Critique of Everyday Life*, 1:207.

58. Neil Smith, foreword to Henri Lefebvre, *The Urban Revolution* (Minneapolis: University of Minnesota Press), xv.

59. For one interesting application of this, see Tom Mels, "Nature, Home, and Scenery: The Official Spatialities of Swedish National Parks," *Environment and Planning D: Society and Space* 20 (2002): 135–54.

60. Neil Smith, "Antinomies of Space and Nature in Henri Lefebvre," in *Philosophy and Geography 2: The Production of Public Space*, ed. Andrew Light and Jonathan M. Smith (Lanham, Md.: Rowman and Littlefield, 1998).

61. Henri Lefebvre, *Introduction to Modernity: Twelve Preludes September 1959–May 1961* (London: Verso, 1995), 132–56.

62. Lefebvre, *Critique of Everyday Life*, 2:64.

63. Lefebvre, *Introduction to Modernity*, 137d.

64. Ibid., 133.

65. Alfred Schmidt, *The Concept of Nature in Marx* (London: New Left Books, 1971). For more on such a position, see Smith, *Uneven Development*, and "Antinomies of Space and Nature in Henri Lefebvre."

66. Ibid.
67. Lefebvre, *Introduction to Modernity*, 146.
68. Ibid., 148.
69. Lefebvre, *Critique of Everyday Life*, 3:168.
70. Ibid., 177.
71. Stuart Elden, "*Mondialisation* before Globalization: Lefebvre and Axelos," in *Space, Difference, Everyday Life: Reading Henri Lefebvre*, ed. Goonewardena, Kipfer, Milgrom, and Schmid, 86.
72. Lefebvre, *Critique of Everyday Life*, 2:8.
73. Ibid., 48–49 (emphasis in original).
74. Henri Lefebvre, *Rhythmanalysis* (London: Continuum, 2000).
75. Smith, "Antinomies of Space and Nature in Henri Lefebvre," 52. All page numbers refer to the English translation of *The Production of Space*.
76. Ibid., 53.
77. Lefebvre, *The Production of Space*, 68. Intriguingly, in this narrower understanding of nature as produced through its apprehension and transformation by the sense organs, we find a conception much closer to Feenberg's reading of Marx as seen in chapter 4.
78. Lefebvre, "Toward a Leftist Cultural Politics," in *Marxism and the Interpretation of Culture*, ed. Nelson and Grossberg, 83.
79. Lefebvre, *Everyday Life in the Modern World*, 2.
80. Lefebvre, *Critique of Everyday Life*, 1:23.
81. Merrifield, *Henri Lefebvre*, 6.
82. Lefebvre, *Critique of Everyday Life*, 2:37.
83. Andrew Feenberg, *Lukács, Marx, and the Sources of Critical Theory* (Oxford: Oxford University Press, 1981), 218.
84. Lefebvre, *Everyday Life in the Modern World*, 204.
85. Roberts, *Philosophizing the Everyday*, 18, 68.
86. Ibid., 69.
87. Miles, *Urban Avant-Gardes*; Rendell, *Art and Architecture*.
88. Roberts, *Philosophizing the Everyday*, 69.
89. Nicolas Bourriaud, *Relational Aesthetics* (Paris: Les Presses du Reel, 2002).
90. Personal communication.
91. Interview with Jim Segers, City Mine(d), June 20, 2006.

Conclusion

1. Quoted in Angelo Quattrocchi and Tom Nairn, *The Beginning of the End: France, May 1968* (London: Panther Books, 1968).
2. David Harvey and Donna Haraway, "Nature, Politics, and Possibilities: A Debate and Discussion with David Harvey and Donna Haraway," *Environment and Planning D: Society and Space* 13 (1995): 5, 507–27.

3. Erik Swyngedouw, "'Circulations and Metabolisms: (Hybrid) Natures and (Cyborg) Cities," *Science as Culture* 15 (2006): 2, 105–21.

4. Quattrocchi and Nairn, *The Beginning of the End*, 7.

5. Henri Lefebvre, *Writings on Cities*, ed. Eleonore Kofman and Elizabeth Lebas (Oxford: Blackwell, 1996), 173.

6. Andy Merrifield, "Guest Editorial: Magical Marxism," *Environment and Planning D: Society and Space* 27 (2009): 386.

Index

Abahlali baseMjondolo, 106
actor network theory, 3, 6, 69, 73; actor network approaches, 5, 14
African National Congress, 96–99, 102–4, 106–7; Youth League, 96–97, 102
alienation, xii–xiii, 24, 26, 28–35, 43, 50, 58, 78, 81, 113–14, 117, 125, 146n25
Althusser, Louis, 26, 58, 144n25, 145n62, 147n34
Amaoti, 21–23, 31–32, 42–43, 45, 75, 96
antihumanism, xxii
apartheid, 71, 75, 78, 88–96, 98–100, 103, 105–6
artistic practice, xi, xx–xxi, 35–36, 38, 40–41, 125
Arvatov, Boris, 117

Badmington, Neil, 15
Barthes, Roland, 117
Benjamin, Walter, xx–xxi, 117, 124
Benton, Ted, xv, 16, 24, 37, 138n14
Berkeley School, 2
Berman, Marshall, 48
Black Local Authorities, 95
Bond, Patrick, 90
Bourriaud, Nicolas, 127
Brecht, Bertolt, xxi, 75, 124

Bukharin, Nikolai, 61, 85, 107; *The Theory of Historical Materialism: A Manual of Popular Sociology*, 61, 85
Burkett, Paul, 25, 47, 64, 65
Business Partners for Development (BPD), 100–103

Carson, Rachel, xv
Castree, Noel, 12–13, 15,141n36, 141n39
cathartic moment, 81
City Mine(d), 109–10, 126–29
climate change, xvi, xvii, 134
coherence (in Gramsci), 79, 81, 139n40
consciousness, x, xiii, xv xxvi, 10, 27, 29, 41, 45–74; imputed, xxiv, 47, 53, 56–61; radical ecological, xxiv, 129
critical spatial practice, xi, xxi, 40, 111, 125–26
Cronon, William, 5
cyborg, xxiv, 45, 47, 56, 60; cyborg city, 68–71, 110, 126; cyborg ecologies, 68; cyborg urbanization, 68

daily life (in Lefebvre), 57, 117
Daniels, Stephen & Endfield, Georgina, xvi
de la Blache, Paul Vidal, 2, 17

determinism, 17, 24, 36; environmental, xiv, 17, 61; technological, 61
dialectics, xviii, 14, 15, 17, 45, 63; dialectical moment, xvii, 59; dialecticism, xi, 114; dialectics of nature, 47, 48, 62, 64
Diamond, Jared, xiv
Durban, 21–22, 71, 73, 87–93, 96, 102, 125

eco-marxist, 8
Elden, Stuart, 111, 113, 114, 122
El Niño, 133
environmentalism, x, 53, 67, 68, 72, 87; environmental movement, xv, 1, 116, 126
environmental justice, x, xxii, 1, 5
Epicurus, xxiii, 25, 27, 34, 38
eThekwini Municipality, 22, 154n94; eThekwini Water Services, 100, 101,154n89
everyday life, ix, xi, xii, xiii, xiv, xvii, xix, xx, xxi, xxiv, xxv, xxvi, 17, 19, 32, 37, 46, 47, 56, 57, 58, 60, 67, 68, 72, 74, 78, 99, 109, 112–19, 122–25, 128–29, 131, 134, 136, 157n49; critique of, ix; xiii, xiv, xx, xxiv, xxv, 37, 46, 47, 56, 109, 112, 115–19, 129, 132, 144n40; everydayness, xiii, 157n40; sensuous creation of, x

Feenberg, Andrew, 27, 36–38, 40, 59, 62–64, 83, 124,158n77
feminist standpoint theory, 46, 68
Feuerbach, Ludwig, xxiii, 26, 27, 28, 38–40, 53, 84, 145n47; *Principles of the Philosophy of the Future*, 28; *The Theses on Feuerbach* (Marx), 26, 27, 39, 40, 80, 84–85, 113, 156n20
Floyd, Kevin, 56

Fontana, Benedetto, 86–87
Foster, John Bellamy, 25–27, 29, 31, 32, 34, 107, 141n42, 143n15, 143n16, 145n53
Frankfurt School, 5, 7–8, 25, 120
Free Basic Water (South Africa), 43, 71

Gandy, Matthew, 6
gendered division of labor, 45, 54, 70, 134
Germinal, 132, 147n38
Glacken, Clarence, 2
Global Environmental Change, xvi–xvii. *See also* climate change; global warming
global warming, xvi, 134
good sense, 81–82, 107
Gormley, Anthony, 128
Gramsci, Antonio, xviii, xx, xxi, xxiv, xxv, 4, 26, 40, 57, 61, 63, 64, 74, 75–87, 95, 99, 107–8, 112, 113, 117, 118, 128 129, 135, 137n2, 139n40, 144n25, 145n62, 148n59, 149n86, 151n1; absolute immanence, absolute historicism, absolute humanism, 79, 104, 151n1; civil society, 106; common sense, 57, 81, 82, 83, 94, 96, 100, 107; concept of nature, 84–87; and Fordism, 104; hegemony, xxv, 78, 85, 99, 104, 105, 106, 113, 117; *Prison Notebooks*, ix, 40, 59, 63, 75, 76, 77, 84, 86, 87, 106, 107; theory of the party, 59, 106
Grundmann, Reiner, 13, 16, 37

Hall, Stuart, 47, 79, 81
Haraway, Donna, x, 1, 15, 17, 21, 23, 24, 26, 33, 55, 67, 68, 69, 72, 132, 145n47; *The Cyborg Manifesto*, 70; encounter value, 16; on the figure of the cyborg, xxiv, 56

Hart, Gillian, 95
Hartsock, Nancy, xviii, xix, 54, 55, 60, 70
Harvey, David, xviii; xx, 2, 3, 4, 5, 13, 14, 17, 31, 32, 33, 37, 68, 111, 113, 142n59; on dialectics, xviii–xx, 17; *Limits to Capital*, 112; on the production of space, 112
Hegel, G. W. F., xi, xii, 3, 7, 26, 27, 28–29, 49, 50, 113, 114, 137n2; alienation, 29; *Aufhebung*, 114; concept of nature, 28; and Haiti (Buck-Morss), 52; Hegelian dialectic, 7, 28, 45, 115, 117, 118; Hegelian Marxism, 49; idealism, 39, 81; method, 46, 50, 51, 73
Heine, Heinrich, xv
Hemson, David, 88, 96, 153n65
Herwegh, Georg, xv
Heynen, Nik, Maria Kaika, and Erik Swyngedouw, 5, 14
Hinchliffe, Steve, 14
Holloway, John, 24, 76
Hughes, Heather, 93–94

idealism, xi, 17, 26, 28, 39; criticisms of idealism in the work of Lukács, 58–59, 62, 64
immanent critique, ix, xii, xiii, xxiv, xxv, 19, 46, 48, 60, 64, 67, 68, 72, 74, 75, 76, 112, 115, 116, 129, 132, 135, 136, 151n1
imputed consciousness, xxiv, 47, 53, 56–61
Inanda, xxv, 21–23, 31, 78, 88–94, 96–100, 102, 104–6, 108; Inanda Civic Association, 97; Inanda Development Forum, 97–98
Inkatha Freedom Party, 22, 88, 94, 98; Inkatha militias, 93

Jameson, Fredric, 55, 58–60, 67
Jay, Martin, 8, 63, 64, 146n28

Kaarsholm, Preben, 96, 98
Kaika, Maria, 6, 14, 22, 41, 70; and Erik Swyngedouw, 4, 5
Kant, Immanuel, 7, 28, 37; Kantian influences on Marx (according to Schmidt), 7, 38, 120; Kantian "thing-in-itself," 63; neo-Kantianism in the early Lukács, 49, 51
Kaplan, Alice, and Kristin Ross, 113, 117
Kipfer, Stefan, 113, 116; Kipfer et al., 111, 112
Kloppenburg, Jack Ralph, 12
KwaMashu, 89

labor: abstract, 30; concrete, 30. *See also* gendered division of labor
Latour, Bruno, 7, 14, 43, 139n37
Lefebvre, Henri, xviii, xix, xx–xxi, xxiii, xxv, 4, 6–7, 8, 25, 56–57, 59, 86, 109–29, 136; conception of nature, xxv, 6, 8, 25, 86, 110, 111, 119–23, 129; critique of everyday life 113, 115–19; everyday life as a work of art, xi, 37, 110, 125, 128; metaphilosophy, ix, 111–12; moment, 110, 112, 114–15, 118; production of space, 6, 110, 112–13
Levins, Richard, and Richard Lewontin, 29
Livingstone, David N., 2
Lorimer, Jamie, 14
Löwy, Michael, 46, 48–50, 58–60
Lukács, Georg, xviii, xxiv, 4, 26, 37, 45–74, 79, 83, 85, 86, 113, 117, 118, 129, 132, 135, 137n2, 141n32, 146n28, 148n59, 149n86; concept of nature, 36, 47–48, 60–74; dialectical approach, 26, 50–53, 60–65; imputed consciousness, xxiv, 47, 53, 56–61, 118; *History and Class Consciousness*, xxiv, 46–60, 62–69,

73; *Lenin*, 50, 58, 60; *Soul and Form*, 58; *Tactics and Ethics*, 49; *Tailism and the Dialectic*, 59, 65–66; *The Theory of the Novel*, 48–49, 52
Luxemburg, Rosa, 50, 60, 82, 146n28

Malthus, Thomas, 24; Malthusian claims, x
Mandela, Nelson, 96
Mann, Thomas, 48
Manquele, Christina, 101
Marx, Karl, ix, x, xii, xiii, xiv, xv, xx, xxiii, xxiv, 3, 4, 15, 49, 50, 51, 53, 54, 62, 67, 71, 73, 77, 80, 81, 84–85, 88, 95, 113–17, 119, 141n39, 144n47, 147n34, 158n77; *Capital*, xi, xii, xix, 9, 17, 26, 112; concept of nature, xv, 7–13, 21–44, 120, 129, 135; *Critique of Hegel's Doctrine of the State*, 116; dialectical approach, xi, xvi–xx, 16–18, 46, 49, 65, 69, 137n2; *Early Writings*, 21, 113; *The Economic and Philosophical Manuscripts* (also referred to as the Paris Manuscripts), 9, 23–29, 34–35, 39, 113, 120, 144n47; *The Eighteenth Brumaire of Louis Bonaparte*, 59; *The German Ideology*, 9, 26, 39; *Grundrisse*, 9, 10, 34; *Introduction to a Critique of Hegel's Philosophy of Right*, 82; *The Theses on Feuerbach*, 26, 27, 39, 40, 80, 84, 85, 113, 116, 156n20
McLellan, David, 35, 144n47
Meer, Fatima, 93
Meillassoux, Claude, 70
Merrifield, Andy, 118, 124, 136
Mészáros, István, 29
metabolism, 5, 6, 21, 25, 29, 31, 43, 68, 133; metabolic process, 6, 7, 9, 12, 34, 71, 135; metabolic rift, 25, 26, 31, 32

Miles, Malcolm, xxi, 40, 41
Mitchell, Don, 1; and Scott Kirsch, 73
Mohanty, Chandra, 72

nature: domination of 8, 9, 25, 34, 37, 86, 87; first and second, 11–12, 30, 121–22; production of, xxii, xxiii, 1–19, 25, 32, 33, 34, 38, 40, 66, 73, 76, 78, 109, 112, 120, 123–25, 135, 141n36; urbanization of, 4–6, 41, 42, 106, 119
Ncwane, Thulani, 96–98, 106, 143n3
Ngceshu, Sabatha, 96
Nietzsche, Friedrich, 114, 115
Nold, Christian, xxiii, 41–43
nonhuman agency, xxii, 73

Ollman, Bertell, 138n19

philosophy of praxis, xii–xiv, xx, xxiv, 4, 26, 29, 34, 57, 61–63, 67, 69, 73, 75–87, 103, 104–8, 113, 116, 117, 125, 128, 129, 135, 145n47
PLATFORM, xxi
posthumanism, 14–16, 18, 46, 73
postpolitical, xvii
Proboscis, 42–43
production of nature, xxii, xxiii, 1–19, 25, 32, 33, 34, 38, 40, 66, 73, 76, 78, 109, 112, 120, 123–25, 135, 141n36
production of space, xxv, 6–7, 25, 40–41, 110–15, 119, 122, 123, 128. *See also* Lefebvre, Henri: production of space

Quattrocchi, Angelo, and Tom Nairn, 132, 135

Rees, John, 65
Rendell, Jane, xxi, 40
Ricardo, David, 30

Roberts, John, xxv, 57–58, 112, 117–18, 125, 128
Roemer, John, xviii
Romantic Movement, xv, 138n14
Roy, Arudhati, x

Sachs, Jeffrey, xiv
Sandoval, Chela, 45, 55, 56, 72
Schmid, Christian, 114–15
Schmidt, Alfred, 5, 7–9, 13, 26, 38, 62, 120
Seneque, Garth, 90, 92
senses, xxiii, 25, 27, 35–42, 46, 145n47; sensuous creation, x; sensuousness, xxiii, 21, 35, 39, 43. *See also* good sense; Gramsci, Antonio: common sense
Shmuely, Andrew, 111
Simmel, Georg, 41, 49
Sitas, Ari, 93, 95
situated knowledges, xxiv, 46, 58, 68, 71
Smith, Neil, xviii, xxiii, 1–19, 25, 27, 30, 31, 34, 38, 48, 76, 110, 112, 119–20, 122–24, 135, 141n32, 141n39; production of nature, xxii, xxiii, 1–19, 25, 32, 33, 34, 38, 40, 66, 73, 76, 78, 109, 112, 120, 123–25, 135, 141n36; *Uneven Development*, 6
socially necessary labour time, 30
social reproduction, xiii, 70–71
Soja, Ed, 111, 112, 114
species being, 18, 32–35, 117

Stedman Jones, Gareth, 58, 59, 62, 148n64
Swyngedouw, Erik, xvii, 5, 6, 14, 68, 71, 133, 139n37. *See also* Heynen, Nik, Maria Kaika, and Erik Swyngedouw; Kaika, Maria

technonatures, 68–71, 110
Thompson, E.P., 142n2

United Democratic Front, 88, 92–94
Urban Foundation (South Africa), 90–91
urban interventions, xi, xxi, 41, 111, 125
urbanization of nature, 4–6, 41, 42, 106, 119
urban political ecology, xxi, xxiii, 3–5, 126; urban political ecologists, xxii, 3, 133

Vaneigem, Raoul, 117
Vivendi Water (Veolia), 100–101
Vogel, Steven, 64–66
von Liebig, Justus, 25

Whatmore, Sarah, 14
White and Wilbert, 68, 72
Wittfogel, Karl, 48
Wolpe, Harold, xiii, 71

Young Hegelians, 28
Young, Robert, 61–62

Žižek, Slavoj, 45, 47, 50, 53, 60, 67, 71, 148n59

Alex Loftus is lecturer in geography at Royal Holloway, University of London.